Trig or Treat

An encyclopedia of
Trigonometric Identity Proofs
(TIPs)

Intellectually challenging games

Trig or

An encyclopedia of
Trigonometric Identity Proofs
(TIPs)

Treat

Intellectually challenging games

Y E O Adrian

M.A., Ph.D. (Cambridge University)
Honorary Fellow, Christ's College, Cambridge University

World Scientific

NEW JERSEY · LONDON · SINGAPORE · BEIJING
SHANGHAI · HONG KONG · TAIPEI · CHENNAI

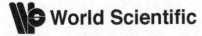 胜利出版私人有限公司

Published by

World Scientific Publishing Co. Pte. Ltd.

5 Toh Tuck Link, Singapore 596224

USA office: 27 Warren Street, Suite 401-402, Hackensack, NJ 07601

UK office: 57 Shelton Street, Covent Garden, London WC2H 9HE

British Library Cataloguing-in-Publication Data
A catalogue record for this book is available from the British Library.

TRIG OR TREAT
An Encyclopedia of Trigonometric Identity Proofs (TIPs) with Intellectually
Challenging Games

ISBN-13 978-981-277-618-1
ISBN-10 981-277-618-4
ISBN-13 978-981-277-619-8 (pbk)
ISBN-10 981-277-619-2 (pbk)

Printed in Singapore.

dedicated to the memory
of my mother

tan peck hiah

$$\tan A + \tan B + \tan C \equiv \tan A \tan B \tan C$$
$$(A + B + C = 180°)$$

$$\tan A + \tan(A + 120°) + \tan(A + 240°) = 3\tan(3A + 360°)$$

$$\tan^{-1}\left(\frac{1}{1}\right) = \tan^{-1}\left(\frac{1}{2}\right) + \tan^{-1}\left(\frac{1}{3}\right)$$

$$\frac{1}{\pi} = \frac{1}{2^2}\tan\frac{\pi}{2^2} + \frac{1}{2^3}\tan\frac{\pi}{2^3} + \frac{1}{2^4}\tan\frac{\pi}{2^4} + \frac{1}{2^5}\tan\frac{\pi}{2^5} + \cdots$$

Contents

Preface

Many students find "Trigonometry" to be a difficult topic in a difficult subject "Math". Yet most students have no difficulty with computer games, and enjoy playing them even though many of these games have lots of pieces to manipulate, and are subjected to complex rules. For example, a simple game like "Tetris" has seven different pieces; and the player has to orientate and manipulate each piece in turn, as it falls. The objective is to construct a solid wall with all the random pieces — and to do it, racing against the clock.

"Trigonometry", or "Trig" for short, can be thought of as an intellectual equivalent of "Tetris". There are six main pieces to manipulate. Three of them, sine, cosine, and tangent, are most important — that is why this branch of Math is called Trigonometry; the "tri" refers to three functions, three angles and three sides of a triangle. And there is only one simple rule — Logic. Trig can be thought of as a game that involves the logical manipulation of various trig pieces to achieve different identities and equations, and to solve numerical problems.

Trig can also be viewed as a non-numerical equivalent of the number game "Sudoku". The logic and the arrangement of the digits 1 to 9, is now applied to the six trig pieces — sine, cosine, tangent, cosecant, secant and cotangent.

Tetris and Sudoku are both simple games that give lots of fun and pleasure. Trig is also a simple game, but with a vital difference — knowledge of it has invaluable applications in Math, surveying, building, navigation, astronomy, and other branches of science, engineering, and technology.

Adults, children and students can play Sudoku and Tetris for hours on end. So they should have little difficulty playing "Trig", if they derive similar fun and pleasure from it.

Albert Einstein said:

> *"Everything should be made as simple as possible,*
> *But not simpler"*.

This book seeks to make Trig as simple as possible, by treating it as a game — albeit, an intellectual game — as interesting and stimulating as Tetris and Sudoku. Mastering of Trig will not only give mental and intellectual satisfaction and pleasure, but it will also lead to beneficial results in one's future career and life.

This book is the third math book* that I have written for my two grandchildren, Kathryn and Rebecca, ages 5 and 7, respectively. The challenge that I set for myself here is to explain Trig so simply that my seven-year-old granddaughter, Rebecca, can understand it. Indeed she has been able to do some of the "Level-One-Games". My hope is that in the coming years both Kathryn and Rebecca would be able to play the "Level-Two-Games" and the "Level-Three-Games" in this book also.

I thank Lim Sook Cheng and her excellent team at World Scientific Publishing for the production of this book; and Zee Jiak Gek for her meticulous reading and critique of all the details in the manuscript.

---

* The other two books are " The Pleasures of Pi,e and Other Interesting Numbers", and "Are You the King or Are You the Joker?".

Pythagoras' Theorem

$$a^2 + b^2 = c^2$$

$$3^2 + 4^2 = 5^2$$

$\text{Sin } A$ $\sin^2 A + \cos^2 A \equiv 1$

Kathryn
Age: 5

12·08·2007

Trigonometry

$\sin A \, \cos A \, \operatorname{cosec} A \, \sec A \equiv 1$

$\mathsf{LHS} = \sin A \, \cos A \, \operatorname{cosec} A \, \sec A$

$\phantom{\mathsf{LHS}} = \sin A \, \cos A \, \dfrac{1}{\sin A} \, \dfrac{1}{\cos A}$

$\phantom{\mathsf{LHS}} = 1$

$\phantom{\mathsf{LHS}} \equiv \mathsf{RHS}$

$\dfrac{\tan A + \cos A}{\sin A} \equiv \sec A + \cot A$

$\mathsf{LHS} = \dfrac{\tan A + \cos A}{\sin A}$

$\phantom{\mathsf{LHS}} = \dfrac{\tan A}{\sin A} + \dfrac{\cos A}{\sin A}$

$\phantom{\mathsf{LHS}} = \dfrac{\sin A}{\cos A} \cdot \dfrac{1}{\sin A} + \dfrac{\cos A}{\sin A}$

$\phantom{\mathsf{LHS}} = \sec A + \cot A$

$\phantom{\mathsf{LHS}} \equiv \mathsf{RHS}$

Rebecca
Age 7
12.08.2007

Introduction

The approach taken in this book is to treat Trig as a game. Beginning with only the definition of **sine**, the superstar of Trig, the book introduces the reader slowly to the basics of Trig. Then, by applying simple logic, the two co-stars, **cosine** and **tangent**, are introduced.

Thereafter three supporting starlets, named the reciprocals — *cosecant*, *secant* and *cotangent* — are added. With these six pieces, the application of simple logic, arithmetic and algebra will give countless Trig equations called identities. Played like jigsaw puzzles, Tetris and Sudoku, moving the Trig pieces around to give different identities can be a lot of fun.

As with other games and puzzles, practice can lead to greater skill and mental agility. About 300 games (proofs) are provided in this book to give fun (and confidence) to readers who want to try their hands (and work their brains) on these intellectual games. The numerous games are broadly grouped into three overlapping levels — Level-One-Games (Easy Proofs), Level-Two-Games, (Less-Easy Proofs) and Level-Three-Games (Not-so-Easy Proofs).

For the first time ever, a "Concordance of Trigonometric Identities" has been created. Trigonometric identities are given a 6-digit code, which enables readers (and students) to have easy reference to the identity to be proved, and to locate rapidly the proof in the Encyclopedia of Trigonometric Identity Proofs (TIPs) in the Appendix.

Readers are welcome to look at the identities in the Concordance first, and try their hand at proving any of the identities, prior to looking at the detailed proofs in the Encyclopedia. (Some identities which may appear simple, may be difficult to prove; conversely, some complex-looking identities may turn out to be relatively easy!)

The games provide the challenge to readers to match their skills, and progress up the ladder of increasing intellectual agility. If you are really good in Trig, then the speed of proving the identities is the speed with which you write out the proofs, i.e. your brain works faster than your brawn (hand).

Have fun with Trig!

Trig — Level One

The Basics of Trigonometry

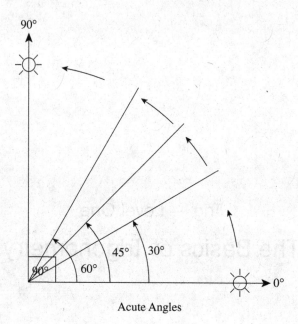

Acute Angles

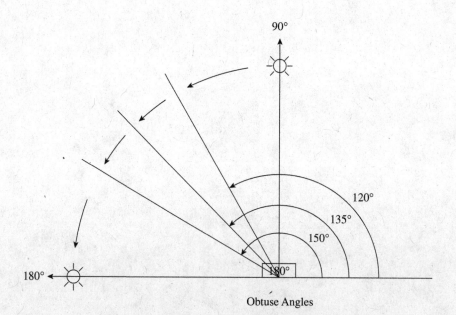

Obtuse Angles

Measuring Angles

"The sun rises in the east, and sets in the west". Similarly, the measuring of angles begins "in the east" (0°), goes counterclockwise, up into the overhead sky at noon (90°) and sets in the west (180°).

People in many ancient civilisations (including the Babylonians, Mesopotamians and the Egyptians) used a numbering system based on 60 called the sexagesimal system. This resulted in the convention of 360° (60° × 6) for the angle round a point. This convention for measuring angles continues to the present day, despite the widespread use of the metric system based on decimals (10's). Another sexagesimal legacy from the past is the use of 60 seconds in a minute, and 60 minutes in an hour.

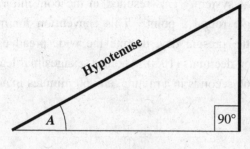

Sine

Hypotenuse

Opposite
side of
angle A

A

90°

$$\text{sine } A = \frac{\textbf{Opposite}}{\textbf{Hypotenuse}}$$

$$\sin A = \frac{O}{H}$$

Sine

Over the centuries, many civilisations used calculations based on right-angled triangles and the relationships of their sides for various purposes, including the building of monuments such as palaces, temples, and pyramids and other tombs for their rulers. Some of these mathematical techniques were also applicable to the study of the stars (astronomy) which led to calender making.

The origins of Trig are lost in the mist of antiquity. One of the earliest recorded reference to the concept of "the sine of an angle" — "*jya*" — was found in a sixth century Indian math book. This word was later translated into "*jiba*" or "*jaib*" in Arabic. A further translation into Latin converted the word into "*sinus*", meaning a bay or curve, the same meaning as "*jaib*". This was further simplified in the 17th century into English — "sine", and abbreviated as "sin" (but always pronounced as "SINE"and not "SIN".)

Sine is simply the name of a specific ratio:

$$\text{sine of an angle } (A) = \frac{\text{length of the opposite side of angle } (A)}{\text{length of the hypotenuse}}$$

This definition is often abbreviated to

$$\sin A = \frac{O}{H}$$

You cannot do Trig if you cannot remember the definition of sin! There are many simple ways of remembering. How about:

1. O/H lang SINE?*
2. O/H, it's so SINple?

Can you create your own mnemonics?

---

*Sounds like "Auld Lang Syne", the universally popular song sung at the stroke of midnight on New Year's Day.

Cosine

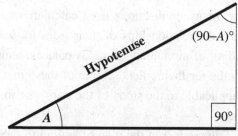

$$\cos A = \frac{\textbf{Adjacent}}{\textbf{Hypotenuse}}$$

$$= \frac{A}{H}$$

Tangent

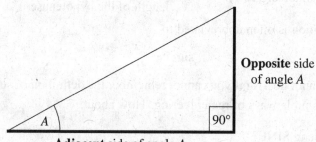

$$\tan A = \frac{\textbf{Opposite}}{\textbf{Adjacent}}$$

$$= \frac{O}{A}$$

$$\tan A = \frac{\sin A}{\cos A}$$

Cosine and Tangent

The complementary angle to the angle A in a right-angled triangle is the third angle, with the value of $(90 - A)°$, because the three angles of a triangle sum to a total of $180°$. The term **"co-sine"** was derived from the phrase "the **sine** of the **complementary angle**"

$$\text{co-sine } A = \text{sine of complementary angle of } A$$

$$= \text{sine}(90 - A)°$$

$$\therefore \cos A = \frac{\text{length of the adjacent side of angle } A}{\text{length of the hypotenuse}}$$

$$= \frac{A}{H}$$

Tangent* is defined as the ratio of $\sin A / \cos A$

$$\therefore \tan A = \frac{\sin A}{\cos A}$$

$$= \frac{\dfrac{O}{H}}{\dfrac{A}{H}}$$

$$= \frac{O}{A}$$

---

*This ratio (tangent) should be distinguished from the line which touches a circle, which is also called tangent in geometry.

Reciprocals

$$\csc A = \frac{1}{\sin A}$$

$$\sec A = \frac{1}{\cos A}$$

$$\cot A = \frac{1}{\tan A}$$

$$\tan A = \frac{\sin A}{\cos A}$$

$$\cot A = \frac{\cos A}{\sin A}$$

Reciprocals

The superstar "sin" and its two co-stars (cos and tan) make up the three key players in Trig. Their definitions and their relationships are essential for all problems in Trig. Hence it is important that they be committed to memory.

Three more trig terms — the supporting cast — are also used. These are known as the reciprocals, and are best remembered as the reciprocals of sin, cos and tan.

$$\frac{1}{\sin A} = \operatorname{cosec} A \quad \text{(cosecant)}$$

$$\frac{1}{\cos A} = \sec A \quad \text{(secant)}$$

$$\frac{1}{\tan A} = \cot A \quad \text{(cotangent)}$$

These reciprocals are rarely used in applications in science, engineering and technology. But for intellectual gymnastics (and in examinations!), these reciprocals are often used in equations and identities.

Pythagoras' Theorem

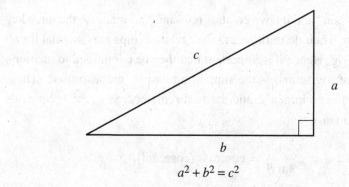

$$a^2 + b^2 = c^2$$

The Famous "3-4-5 Triangle"

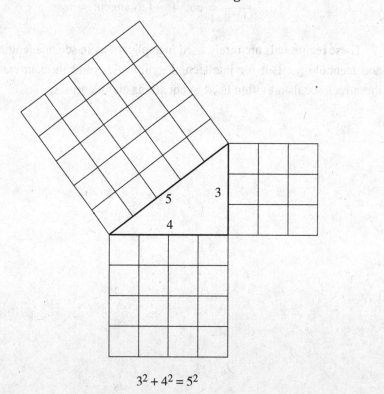

$$3^2 + 4^2 = 5^2$$

Pythagoras' Theorem*

The most well-known theorem in Math, which practically every student has learnt, is the Pythagoras' Theorem, named after the Greek mathematician Pythagoras ($\sim$580–500 BC).

This theorem states that in a right-angled triangle, the square of the hypotenuse (the longest side) is equal to the sum of the squares of the two other sides. This lengthy statement can be represented accurately in mathematical terms:

$$a^2 + b^2 = c^2$$

where a and b are lengths of the two sides, and c is the length of the hypotenuse, the side facing the right-angle.

The most famous right-angled triangle is the "*3-4-5* triangle":

$$3^2 + 4^2 = 5^2$$
$$(9 + 16 = 25)$$

A less famous sister is the "*5-12-13* triangle" (*5^2 + 12^2 = 13^2; 25 + 144 = 169*).

Recent research has shown that many civilisations, including the Babylonian, the Egyptian, the Chinese and the Indian civilisations, independently knew about the relationship between the squares of the three sides of the right-angled triangle, in some cases, centuries before Pythagoras was born. (This illustrates a truism in Math, that often, your discoveries based on your own efforts, may have been preceded by others. However this does not diminish in any way, the pleasure, excitement and sense of achievement that you experienced — the so-called "eureka effect". Indeed, it proves that you have a mathematical mind, capable of the same deep thoughts as the ancient heroes of Math.)

---

*A theorem is simply a mathematical statement whose validity has been proven by meticulous mathematical reasoning.

· Trig Equivalent of Pythagoras' Theorem

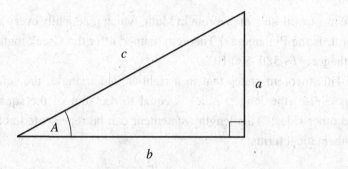

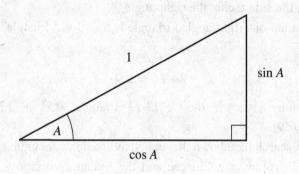

$$\sin^2 A + \cos^2 A = 1$$

$$\sin^2 A = 1 - \cos^2 A$$

$$\cos^2 A = 1 - \sin^2 A$$

Trig Equivalent of Pythagoras' Theorem

One of the most important of Trig identities* is the trig equivalent of the Pythagoras' Theorem. The proof is simple:

$$\sin A = \frac{a}{c} \qquad \cos A = \frac{b}{c}$$

$$\sin^2 A + \cos^2 A = \frac{a^2}{c^2} + \frac{b^2}{c^2}$$

$$= \frac{a^2 + b^2}{c^2}$$

$$= \frac{c^2}{c^2} \quad \left(\begin{array}{c} \text{by Pythagoras' Theorem} \\ a^2 + b^2 = c^2 \end{array} \right)$$

$$= 1$$

$$\boxed{\sin^2 A + \cos^2 A \equiv 1}$$

A simpler visual proof can be obtained by using a special right-angled triangle with a hypotenuse of unit length (1).

Then the length of the opposite side is now equal to $\sin A$, and the length of the adjacent side is equal to $\cos A$ (see figure opposite). Then by Pythagoras' Theorem:

$$\sin^2 A + \cos^2 A \equiv 1^2$$

$$\sin^2 A + \cos^2 A \equiv 1.$$

This unity trig identity is the simplest and the most important of all trig identities. It is also extremely useful in helping to solve trig problems. Whenever you see $\sin^2 A$ or $\cos^2 A$, always consider the possibility of using this identity to simplify further.

---

*An identity is a mathematical equation that is true for all values of the angle A. It does not matter whether $A = 30°$, $60°$, $90°$ etc, whether it is acute or obtuse, etc. The symbol ($\equiv$) is used to show that the two sides of an equation are identical.

We can also derive two other identities:

<table>
<tr><td>dividing
by $\sin^2 A$</td><td>$$\frac{\sin^2 A}{\sin^2 A} + \frac{\cos^2 A}{\sin^2 A} \equiv \frac{1}{\sin^2 A}$$
$$1 + \cot^2 A \equiv \operatorname{cosec}^2 A$$</td></tr>
<tr><td>dividing
by $\cos^2 A$</td><td>$$\frac{\sin^2 A}{\cos^2 A} + \frac{\cos^2 A}{\cos^2 A} \equiv \frac{1}{\cos^2 A}$$
$$\tan^2 A + 1 \equiv \sec^2 A.$$</td></tr>
</table>

After this simple introduction, you are now ready to play Level-One-Games (Easy Proofs), some of which seven-year-old Rebecca could play.

The general approach for playing the games (proving the identities) is to:

1. start with the more complex side of the identity (usually the left hand side (LHS));
2. eyeball the key terms, and think in terms of sin and cos of the angle;
3. engage in some mental gymnastics — rearranging and simplifying;
4. whilst at all times, keeping the terms in the right hand side — the final objective — in mind.

Like a guided missile, your logic and math manipulation of the LHS should lead you to zoom in to the RHS.

Have Fun!

Trig — Level Two

Compound Angles

Sine of (the Sum of Two Angles)

$$\sin(A+B) = \sin A \cos B + \cos A \sin B$$

Substituting $B = A$,

$$\sin(A+A) = \sin A \cos A + \cos A \sin A$$

$$\therefore \ \sin 2A = 2 \sin A \cos A \ \text{(Double Angle Formula)}$$

Substituting A with $(A/2)$,

$$\sin 2 \left(\frac{A}{2} \right) = 2 \sin \frac{A}{2} \cos \frac{A}{2}$$

$$\therefore \ \sin A = 2 \sin \frac{A}{2} \cos \frac{A}{2} \ \text{(Half Angle Formula)}$$

Sine of (the Difference of Two Angles)

Substituting B in the first equation by $(-B)$,

$$\sin(A + (-B)) = \sin A \cos(-B) + \cos A \sin(-B)$$

$$\therefore \sin(A - B) = \sin A \cos B - \cos A \sin B$$

since $\cos(-B) = \cos B$

and $\sin(-B) = -\sin B$

See Proof in Addenda p. 395

Sine of the Sum and the Difference of Angles

This is the beginning of Level-Two-Games. The sin of compound angles (i.e. angles that are the sum or the difference of two other angles) can be expressed in terms of trig functions of the single angles.

For example, sin of $(A + B)$ can be expressed as a combination of the sum and products of the sin and cos of A and B separately. Similarly, since we know that $\sin(-A) = -\sin A$, and $\cos(-A) = \cos A$, we can derive $\sin(A - B)$. By judicious substitution (using $B = A$), $\sin(A + B)$ can be changed to $\sin 2A$ and then into $\sin A$ (using $A = 2(A/2)$).

In earlier days, before calculators and computers were available, knowledge of the trig functions of compound angles was invaluable in practical workplace calculations. Knowing the basics for $0°$, $30°$, $45°$, $60°$ and $90°$, one could work out the values of trig functions of $15°$ and $22.5°$ and other such angles by way of these functions. In those days, Trig was both "pure math" and "applied math", useful in many professions involving science, engineering and architecture.

Today, where the pressing of a few buttons on a calculator or computer will give answers for all such calculations, Trig is largely "pure math" — a mental pursuit, an intellectual game.

But what a beautiful game it is, especially when you are immersed in the proving of the vast number of identities!

Cosine of (the Sum of Two Angles)

$$\cos(A+B) = \cos A \cos B - \sin A \sin B$$

> *Note the minus sign on the RHS, even though the sign on the LHS is plus*

Substituting $B = A$,

$$\cos 2A = \cos^2 A - \sin^2 A \ \text{(Double Angle Formula)}$$

$$= 2\cos^2 A - 1$$

$$= 1 - 2\sin^2 A$$

> *Remember:-*
> $s^2 + c^2 = 1$
> $\therefore -s^2 = c^2 - 1$
> *and* $c^2 = 1 - s^2$

Substituting A with $(A/2)$,

$$\cos 2\left(\frac{A}{2}\right) = \cos^2 \frac{A}{2} - \sin^2 \frac{A}{2}$$

$$\therefore \cos A = \cos^2 \frac{A}{2} - \sin^2 \frac{A}{2} \ \text{(Half Angle Formula)}$$

$$= 2\cos^2 \frac{A}{2} - 1$$

$$= 1 - 2\sin^2 \frac{A}{2}$$

Cosine of (the Difference of Two Angles)

Substituting B in the first equation by $(-B)$,

$$\cos(A + (-B)) = \cos A \cos(-B) - \sin A \sin(-B)$$

$$\therefore \ \cos(A - B) = \cos A \cos B + \sin A \sin B$$

since $\cos(-B) = \cos B$

and $\sin(-B) = -\sin B$

> *Note the plus sign on the RHS, even though the sign on the LHS is minus*

See Proof in Addenda p. 397

Cosine of the Sum and Difference of Angles

Cos functions require special attention as occassionally they act in a contrarian, counter-intuitive manner — this arises largely from the fact that $\cos(-A) = \cos A$.

This is the first function where such counter-intuitive behaviour of cos shows itself.

Although the LHS of the cos function is for the *sum* of two angles, the RHS shows a difference of the two products. Students are usually careless and are not sufficiently sensitive to such minor (???) intricacies in Math. Unfortunately such minor (???) inattention can be very costly in examinations because they lead to wrong answers and major (???) losses in marks!

Tangent of (the sum of Two Angles)

$$\tan(A+B) = \frac{\tan A + \tan B}{1 - \tan A \tan B}$$

Note the minus sign in the denominator on the RHS, even though the LHS has a plus sign!

Substituting $B = A$,

$$\tan 2A = \frac{2 \tan A}{1 - \tan^2 A} \quad \text{(Double Angle Forumla)}$$

Substituting A with $(A/2)$,

$$\tan 2\left(\frac{A}{2}\right) = \frac{2 \tan \dfrac{A}{2}}{1 - \tan^2 \dfrac{A}{2}} \tag{1}$$

$$\therefore \ \tan A = \frac{2 \tan \dfrac{A}{2}}{1 - \tan^2 \dfrac{A}{2}} \quad \text{(Half Angle Formula)} \tag{2}$$

Tangent of (the Difference of Two Angles)

Substituting B in the first equation by $(-B)$,

$$\tan(A + (-B)) = \frac{\tan A + \tan(-B)}{1 - \tan A \tan(-B)}$$

Note the plus sign in the denominator on the RHS, even though the LHS has a minus sign!

$$\therefore \tan(A - B) = \frac{\tan A - \tan B}{1 + \tan A \tan B}$$

since $\tan(-B) = -\tan B$

See Proof in Addenda p. 398

Tangent of the Sum and the Difference of Two Angles

The tan formulas for the sum and difference of two angles derive directly from the sin and cos formulas. The double angle formula, $\tan(2A)$, and the half angle formula, $\tan A$, have proven to be extremely valuable in many mathematical proofs, and have resulted in sophisticated methods for the calculation of π to a large number of decimal places. (Would you believe that π has been calculated to 1.24 trillion decimal places — yes, 1,240,000,000,000 decimals?)

Trig or Treat

Summary of Trig Functions for Compound Angles

$$\sin(A+B) = \sin A \cos B + \cos A \sin B$$

$$\sin 2A = 2 \sin A \cos A$$

$$\sin A = 2 \sin \frac{A}{2} \cos \frac{A}{2}$$

$$\sin(A-B) = \sin A \cos B - \cos A \sin B$$

$$\cos(A+B) = \cos A \cos B - \sin A \sin B$$

$$\cos 2A = \cos^2 A - \sin^2 A$$

$$\cos A = \cos^2 \frac{A}{2} - \sin^2 \frac{A}{2}$$

$$\cos(A-B) = \cos A \cos B + \sin A \sin B$$

$$\tan(A+B) = \frac{\tan A + \tan B}{1 - \tan A \tan B}$$

$$\tan 2A = \frac{2 \tan A}{1 - \tan^2 A}$$

$$\tan A = \frac{2 \tan \frac{A}{2}}{1 - \tan^2 \frac{A}{2}}$$

$$\tan(A-B) = \frac{\tan A - \tan B}{1 + \tan A \tan B}$$

The sin, cos and tan of compound angles, and their "double angle" and "half angle" formulas provide the basis for many of the Level-Two-Games. Together with the most important identity:

$$\sin^2 A + \cos^2 A \equiv 1,$$

the 12 identities on the opposite page, make up the total of the key Trig functions. Most other Trig identities can be derived from these "12 + 1" key identities.

The typical student (or Trig player) is expected to know these "12 + 1" key functions (very lucky if you know them instinctively, and very unlucky if you don't). With these "12 + 1" key functions, Level-Two-Games (Less-Easy Proofs) should prove to be easy also.

Within Level-Two-Games, the proofs begin with the simpler ones and progress upwards in difficulty.

Have Fun!

Trig — Level Three

Angles in a Triangle

Special Trig Identities for All
Three Angles in a Triangle

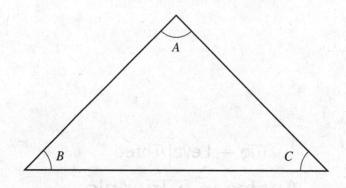

$$A + B + C = 180°$$
$$C = 180° - (A + B)$$
$$\sin C = \sin (180° - (A + B))$$
$$= \sin (A + B)$$
$$\cos C = \cos (180° - (A + B))$$
$$= -\cos (A + B)$$

$$\frac{A}{2} + \frac{B}{2} + \frac{C}{2} = 90°$$
$$\sin \frac{C}{2} = \sin \left(90° - \left(\frac{A + B}{2}\right)\right)$$
$$= \cos \left(\frac{A + B}{2}\right)$$
$$\cos \frac{C}{2} = \cos \left(90° - \left(\frac{A + B}{2}\right)\right)$$
$$= \sin \frac{A + B}{2}$$

$$\sin (A + B + C) = \sin 180° = 0$$
$$\cos (A + B + C) = \cos 180° = -1$$

See Graphs on p. 42

Trig Identities Involving All Three Angles in a Triangle

Some special trig identities apply only when all three angles in a triangle are involved. For such identities, the additional constraint of

$$(A + B + C) = 180°$$

is a critical one.

For such identities the relationship between the three angles is always necessary for simplification, and sometimes result in beautiful identities as seen in some of the examples in Level-Three-Games.

The Sum and Difference of Sine Functions

$$\sin(A+B) = \sin A \cos B + \cos A \sin B$$

$$\sin(A-B) = \sin A \cos B - \cos A \sin B$$

Adding the two equations:

$$\sin(A+B) + \sin(A-B) = 2\sin A \cos B$$

$$\sin S + \sin T = 2\sin\left(\frac{S+T}{2}\right)\cos\left(\frac{S-T}{2}\right)$$

$$Let\ (A+B) = S$$
$$and\ (A-B) = T$$
$$\therefore A = \frac{S+T}{2}$$
$$and\ B = \frac{S-T}{2}$$

Subtracting the second equation from the first:

$$\sin(A+B) - \sin(A-B) = 2\cos A \sin B$$

$$\sin S - \sin T = 2\cos\left(\frac{S+T}{2}\right)\sin\left(\frac{S-T}{2}\right)$$

The Sum and Difference of Sine Functions

Often in Math, the addition or subtraction of similar equations, or the re-arrangement of sums and differences can lead to new insights and new equations which may be of special value.

The equations in the earlier pages are the sin and cos formulas of compound angles.

Here we are looking at the sum and difference of the trig functions of such compound angles, and after further simplification, we derive new relationships. Knowing these new relationships provide greater flexibility and agility in the manipulation of the trig building blocks, and enables new equations or identities to be proved.

The Sum and Difference of Cosine Functions

$$\cos(A+B) = \cos A \cos B - \sin A \sin B$$

$$\cos(A-B) = \cos A \cos B + \sin A \sin B$$

Adding the two equations:

$$\cos(A+B) + \cos(A-B) = 2\cos A \cos B$$

$$\cos S + \cos T = 2\cos\left(\frac{S+T}{2}\right)\cos\left(\frac{S-T}{2}\right)$$

> *Let* $(A+B) = S$
>
> *and* $(A-B) = T$
>
> $$\therefore A = \frac{S+T}{2}$$
>
> *and* $B = \dfrac{S-T}{2}$

Subtracting the second equation from the first:

$$\cos(A+B) - \cos(A-B) = -2\sin A \sin B$$

$$\cos S - \cos T = -2\sin\left(\frac{S+T}{2}\right)\sin\left(\frac{S-T}{2}\right)$$

> *Note the minus sign in front of the RHS terms*

The Sum and Difference of Cosine Functions

Similar addition and subtraction of the cos functions for compound angles give similar equations for the sum and difference of cos functions.

While these functions were extremely useful before calculators and computers were available, they have fallen into disuse in modern times except for purposes of "examinations", to test the student's versatility.

You are now ready to play Level-Three-Games — the "Not-So-Easy" Proofs.*

Have Fun!

---

*There are no simple sum and difference formulas for tan!

Practical Trig

Numerical Values of Special Angles

Numerical Values of Special Angles

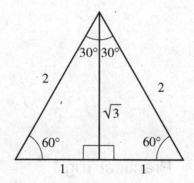

$$\sin 30° = \frac{1}{2} \qquad \cos 30° = \frac{\sqrt{3}}{2} \qquad \tan 30° = \frac{1}{\sqrt{3}}$$

$$\sin 60° = \frac{\sqrt{3}}{2} \qquad \cos 60° = \frac{1}{2} \qquad \tan 60° = \sqrt{3}$$

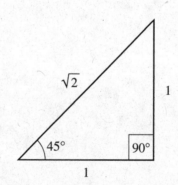

$$\sin 45° = \frac{1}{\sqrt{2}} \qquad \cos 45° = \frac{1}{\sqrt{2}} \qquad \tan 45° = \frac{1}{1}$$

Numerical Values of Special Angles

Five angles are of special interest in trigonometry — 0°, 30°, 45°, 60°, 90°. Therefore it is important to know them and to remember them. The proofs are visual (resulting from the Pythagoras' Theorem), and are easy to follow (see figures on the opposite page). Written in the form of square roots, the ratios are easy to remember, beginning with $\sin 0° = 0$, and $\cos 0° = 1$.

$\sin 0° = 0 \quad = \sqrt{\dfrac{0}{4}}$	$\cos 0° = 1 \quad = \sqrt{\dfrac{4}{4}}$	$\tan 0° = 0$
$\sin 30° = \dfrac{1}{2} \quad = \sqrt{\dfrac{1}{4}}$	$\cos 30° = \dfrac{\sqrt{3}}{2} = \sqrt{\dfrac{3}{4}}$	$\tan 30° = \dfrac{1}{\sqrt{3}}$
$\sin 45° = \dfrac{\sqrt{2}}{2} = \sqrt{\dfrac{2}{4}}$	$\cos 45° = \dfrac{\sqrt{2}}{2} = \sqrt{\dfrac{2}{4}}$	$\tan 45° = 1$
$\sin 60° = \dfrac{\sqrt{3}}{2} = \sqrt{\dfrac{3}{4}}$	$\cos 60° = \dfrac{1}{2} \quad = \sqrt{\dfrac{1}{4}}$	$\tan 45° = \sqrt{3}$
$\sin 90° = 1 \quad = \sqrt{\dfrac{4}{4}}$	$\cos 90° = 0 \quad = \sqrt{\dfrac{0}{4}}$	$\tan 90°$ (indeterminate)

Both $\sin 0°$ and $\tan 0°$ are zero because the length for "opposite" is zero. $\cos 0° = 1$ because the "adjacent" and the hypotenuse are identical.

Similarly $\sin 90° = 1$ because the "opposite" is coincident with the hypotenuse; and $\cos 90° = 0$ because the length of "adjacent" is zero.

It is important to stress than $\tan 90°$ does not have a value (indeterminate); in Math, we refer to it as "tending to infinity" and write it as "$\to \infty$".

All Values of Sin (0° − 360°)

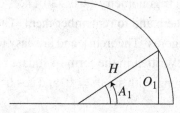

First Quadrant

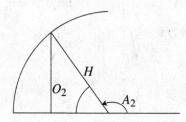

Second Quadrant

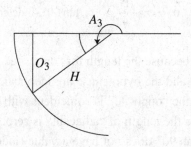

Third Quadrant

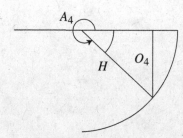

Fourth Quadrant

Values of Obtuse and Negative Angles

Sine

One of the potential obstacles to having fun with Trig games is discovering that not all trig functions are positive. Trig functions of angles larger than 90° (obtuse angles) can sometime have negative values.

One of the easiest ways to overcome these potential obstacles to have fun with Trig is to focus on the values of the sin function. Consider four angles A_1, A_2, A_3 and A_4 in the four quadrants, respectively:

$$\text{First Quadrant:} \quad \sin A_1 = \frac{O_1}{H^*} = \frac{+\text{ve}}{+\text{ve}} = +\text{ve}$$

$$\text{Second Quadrant:} \quad \sin A_2 = \frac{O_2}{H} = \frac{+\text{ve}}{+\text{ve}} = +\text{ve}$$

$$\text{Third Quadrant:} \quad \sin A_3 = \frac{O_3}{H} = \frac{-\text{ve}}{+\text{ve}} = -\text{ve}$$

$$\text{Fourth Quadrant:} \quad \sin A_4 = \frac{O_4}{H} = \frac{-\text{ve}}{+\text{ve}} = -\text{ve}.$$

---

*The hypotenuse always has a positive value.

Value of Sine in the Four Quadrants

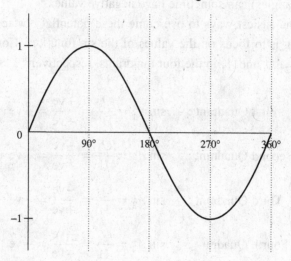

Value of Sine in the Four Quadrants

The values of the sine of angles in the first and second quadrants (i.e. between 0° and 180°) are always positive as seen in the graph on the opposite page, rising from 0 for sin 0° to a maximum value of 1 for sin 90°.

Similarly the values of the sine of angles in the third and fourth quadrants (i.e. between 180° and 360°) are all negative, going to a minimum of −1 for sin 270°, and returning to 0 for sine 360°.

The brain remembers pictures better than equations or words; so commit the graph on the opposite page to memory; and this would prove to be extremely valuable in solving trig problems. This graph is well-known in Math as "the sinusoidal curve" or the "sine curve", for short.

A quick sketch of the sinusoidal curve (done in 10 seconds) will provide a good guide for ensuring that the correct values of sine in the different quadrants are obtained.

It is also important to remember that the sine of a negative angle is the negative of the sine of the angle.

$$\sin(-A) = -\sin A.$$

Value of Cosine in the Four Quadrants

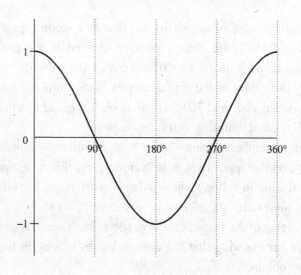

Value of Tangent in the Four Quadrants

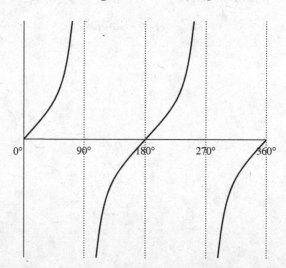

Cosine and Tangent

By similar consideration, the values of the cos and tan of angles in the four quadrants can be obtained. Again it is easier to remember the pictures (graphs on the opposite page).

Always remember that cos 0° = 1, so the graph for cos always begins at 1, and go to −1 for cos 180°. (The cos graph for 0° to 360° looks like a hole in the ground).

Again, remember:

$$\cos(-A) = \cos A.$$

The cos function has this unusual feature and hence cos functions need special attention (i.e. be extra careful with cos functions).

For tan, the value goes from tan 0° = 0 all the way to the indeterminate value for tan 90°. Interestingly, the third quadrant is an exact replica of the first quadrant, and the fourth an exact replica of the second.

Again, remember:

$$\tan(-A) = -\tan A$$

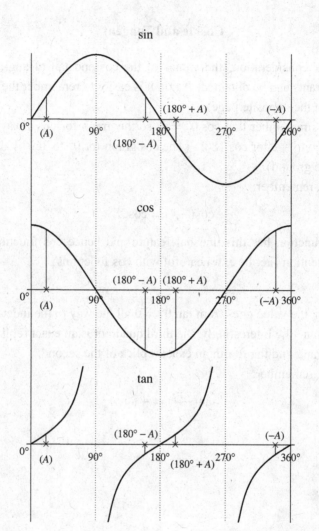

It is very difficult to remember all the equations for calculating the different values of sin, cos and tan of angles in all four quadrants, as well as the negative angles.

Fortunately, the three pictures (graphs on the opposite page) enable easy derivation of their values based on their relation to their primary values in the first quadrant). Let us use an arbitrary angle (say 39°) to see how we can work out the correct values for angles in all the four quadrants.

$$\sin(-39°) = -\sin 39° \qquad\qquad \cos(-39°) = \cos 39°$$

$$2^{nd}\ Q \qquad \sin 141° = \sin(180° - 39°) \qquad\qquad \cos 141° = -\cos(180° - 39°)$$

$$= \sin 39° \qquad\qquad\qquad = -\cos 39°$$

$$3^{rd}\ Q \qquad \sin(180° + 39°) = -\sin 39° \qquad \cos(180° + 39°) = -\cos 39°$$

$$4^{th}\ Q \qquad \sin(360° - 39°) = -\sin 39° \qquad \cos(360° - 39°) = \cos 39°$$

$$\tan(-39°) = -\tan 39°$$

$$2^{nd}\ Q \qquad \tan 141° = -\tan(180° - 39°)$$

$$= -\tan 39°$$

$$3^{rd}\ Q \qquad \tan(180° + 39°) = \tan 39°$$

$$4^{th}\ Q \qquad \tan(360° - 39°) = -\tan 39°$$

Appendix

The Concordance
of
Trigonometric Identities

The six-digit code for the Concordance is based on the number of trig functions on the LHS of the identity e.g. 123 000 means that the LHS has 1 sin, 2 cos, 3 tan and no cosec, sec and cot functions. On the rare occasion when you cannot find the identity in the Concordance, use the functions on the RHS to determine the code.

sin	cos	tan	cosec	sec	cot		Level	Page
0	0	0	0	0	2	$2\cot A\cot 2A \equiv \cot^2 A - 1$	2	288
						$(1+\cot A)^2 + (1-\cot A)^2 \equiv \dfrac{2}{\sin^2 A}$	1	233
						$\cot^4 A + \cot^2 A \equiv \operatorname{cosec}^4 A - \operatorname{cosec}^2 A$	1	188
						$\dfrac{\cot A + 1}{\cot A - 1} \equiv \dfrac{1+\tan A}{1-\tan A}$	1	110
						$\dfrac{1-\cot^2 A}{1+\cot^2 A} \equiv \sin^2 A - \cos^2 A$	1	183
						$\dfrac{\cot^2 A - 1}{2\cot A} \equiv \cot 2A$	2	315
0	0	0	0	0	4	$\dfrac{\cot A\cot B - 1}{\cot A + \cot B} \equiv \cot(A+B)$	2	323
						$\dfrac{\cot A\cot B + 1}{\cot A - \cot B} \equiv \cot(A-B)$	2	274
0	0	0	0	1	0	$2\sec^2 A - 1 \equiv 1 + 2\tan^2 A$	1	201
						$1 + 2\sec^2 A \equiv 2\tan^2 A + 3$	1	128
0	0	0	0	1	1	$\cot A(\sec^2 A - 1) \equiv \tan A$	1	148
0	0	0	0	2	0	$\sec^4 - \sec^2 A \equiv \tan^4 A + \tan^2 A$	1	157
						$\dfrac{\sec A}{1 + \sec A} \equiv \dfrac{1 - \cos A}{\sin^2 A}$	1	203
						$\dfrac{1 + \sec A}{1 - \sec A} \equiv \dfrac{\cos A + 1}{\cos A - 1}$	1	79

sin	cos	tan	cosec	sec	cot		Level	Page
0	0	0	0	2	0	$\dfrac{\sec^2 A}{2 - \sec^2 A} \equiv \sec 2A$	2	295
						$\dfrac{1 + \sec A}{\sec A} \equiv \dfrac{\sin^2 A}{1 - \cos A}$	1	244
						$\dfrac{\sec A + 1}{\sec A - 1} \equiv \cot^2 \dfrac{A}{2}$	2	312
0	0	0	1	0	1	$\csc^2 A - \cot^2 A \equiv 1$	1	145
						$\csc A - \cot A \equiv \dfrac{\sin A}{1 + \cos A}$	1	219
						$\csc^4 A - \cot^4 A \equiv \csc^2 A + \cot^2 A$	1	103
						$\csc 2A - \cot 2A \equiv \tan A$	2	299
						$\dfrac{\csc A - 1}{\cot A} \equiv \dfrac{\cot A}{\csc A + 1}$	1	225
0	0	0	1	1	0	$\csc A \sec A \equiv 2 \csc 2A$	2	270
						$\csc A \sec A \equiv \tan A + \cot A$	1	135
						$\csc^2 A + \sec^2 A \equiv \csc^2 A \sec^2 A$	1	186
						$\dfrac{1}{\sec^2 A} + \dfrac{1}{\csc^2 A} \equiv 1$	1	97
0	0	0	2	0	0	$\dfrac{\csc A - 1}{\csc A + 1} \equiv \dfrac{1 - \sin A}{1 + \sin A}$	1	109
						$\dfrac{\csc A}{1 + \csc A} \equiv \dfrac{1 - \sin A}{\cos^2 A}$	1	215
						$\csc^4 A - \csc^2 A \equiv \cot^4 A + \cot^2 A$	1	158

sin	cos	tan	cosec	sec	cot		Level	Page
0	0	0	2	0	2	$\dfrac{\operatorname{cosec} A \operatorname{cosec} B}{\cos A \cot B - 1} \equiv \sec(A+B)$	2	273
0	0	0	2	2	0	$\dfrac{\sec A}{\operatorname{cosec}^2 A} - \dfrac{\operatorname{cosec} A}{\sec^2 A}$		
						$\equiv (1 + \cot A + \tan A)(\sin A - \cos A)$	1	243
						$\dfrac{\sec A - \operatorname{cosec} A}{\sec A \operatorname{cosec} A} \equiv \sin A - \cos A$	1	113
0	0	1	0	0	0	$\tan^2 A + 1 \equiv \sec^2 A$	1	146
						$\tan\left(45° + \dfrac{A}{2}\right) \equiv \tan A + \sec A$	2	300
						$\tan\left(A - \dfrac{\pi}{4}\right) \equiv \dfrac{\tan A - 1}{\tan A + 1}$	2	285
0	0	1	0	0	1	$\tan A + \cot A \equiv 2\operatorname{cosec} 2A$	2	306
						$\tan A + \cot A \equiv \operatorname{cosec} A \sec A$	1	144
						$\tan^2 A + \cot^2 A + 2 \equiv \operatorname{cosec}^2 A \sec^2 A$	1	101
						$\tan A + \cot A \equiv \dfrac{2}{\sin 2A}$	2	290
						$\dfrac{1 + \cot A}{1 + \tan A} \equiv \cot A$	1	152
						$\dfrac{2}{\cot A \tan 2A} \equiv 1 - \tan^2 A$	2	252
						$\dfrac{1}{\tan A + \cot A} \equiv \dfrac{\sin A}{\sec A}$	1	163
						$\dfrac{1}{2}(\cot A - \tan A) \equiv \cot 2A$	2	269

sin	cos	tan	cosec	sec	cot		Level	Page
0	0	1	0	0	1	$\dfrac{1}{\tan A + \cot A} \equiv \sin A \cos A$	1	166
0	0	1	0	0	2	$\dfrac{\cot A(1+\tan^2 A)}{1+\cot^2 A} \equiv \tan A$	1	180
0	0	1	0	1	0	$\sec 2A - \tan 2A \equiv \tan(45° - A)$	2	303
						$\sec A - \tan A \equiv \dfrac{\cos A}{1+\sin A}$	1	218
						$\sec^4 A - \tan^4 A \equiv \dfrac{1+\sin^2 A}{\cos^2 A}$	1	236
						$(\sec A - \tan A)^2 \equiv \dfrac{1-\sin A}{1+\sin A}$	1	230
						$\dfrac{\tan A}{\sec A - 1} \equiv \dfrac{\sec A + 1}{\tan A}$	1	222
0	0	1	1	0	0	$\tan A \, \text{cosec}\, A \equiv \sec A$	1	104
						$\tan^2 A(\text{cosec}^2 A - 1) \equiv 1$	1	90
0	0	1	1	0	1	$\text{cosec}\, A + \cot A + \tan A \equiv \dfrac{1+\cos A}{\sin A \cos A}$	1	131
						$\dfrac{\text{cosec}\, A}{\cot A + \tan A} \equiv \cos A$	1	159
0	0	1	1	1	0	$\dfrac{\sec A + \text{cosec}\, A}{1+\tan A} \equiv \text{cosec}\, A$	1	162
0	0	1	1	1	1	$(\tan A - \text{cosec}\, A)^2 - (\cot A - \sec A)^2$ $\equiv 2(\text{cosec}\, A - \sec A)$	2	338
0	0	2	0	0	0	$\tan A + \tan 2A \equiv \dfrac{\sin A(4\cos^2 A - 1)}{\cos A \cos 2A}$	2	337

sin	cos	tan	cosec	sec	cot		Level	Page
0	0	2	0	0	0	$\tan(45° + A)\tan(45° - A)$		
						$\equiv \cot(45° + A)\cot(45° - A)$	2	326
						$(1 + \tan A)^2 + (1 - \tan A)^2 \equiv 2\sec^2 A$	1	185
						$\dfrac{1 - \tan^2 A}{1 + \tan^2 A} + 1 \equiv 2\cos^2 A$	1	247
						$\dfrac{1 - \tan^2 A}{1 + \tan^2 A} \equiv \cos 2A$	2	317
						$\dfrac{2\tan A}{1 + \tan^2 A} \equiv \sin 2A$	2	316
						$\dfrac{1 + \tan A}{1 - \tan A} \equiv \dfrac{\cot A + 1}{\cot A - 1}$	1	110
0	0	2	0	0	1	$\dfrac{\tan A(1 + \cot^2 A)}{1 + \tan^2 A} \equiv \cot A$	1	178
						$\tan A(\cot A + \tan A) \equiv \sec^2 A$	1	89
0	0	2	0	0	2	$(\tan A + \cot A)^2 - (\tan A - \cot A)^2 \equiv 4$	1	182
						$\dfrac{\tan A + \tan B}{\cot A + \cot B} \equiv \tan A \tan B$	1	171
						$\dfrac{\tan A - \cot A}{\tan A + \cot A} \equiv \sin^2 A - \cos^2 A$	1	189
						$\dfrac{\cot A - \tan A}{\cot A + \tan A} \equiv \cos 2A$	2	280
						$\dfrac{\tan A - \cot A}{\tan A + \cot A} + 1 \equiv 2\sin^2 A$	1	239
						$\dfrac{\tan A - \cot A}{\tan A + \cot A} \equiv 1 - 2\cos^2 A$	1	245

sin	cos	tan	cosec	sec	cot		Level	Page
0	0	2	0	0	2	$\dfrac{\cot A}{1-\tan A} + \dfrac{\tan A}{1-\cot A} \equiv 1+\tan A+\cot A$	1	224
						$\dfrac{\cot A}{1-\tan A} + \dfrac{\tan A}{1-\cot A} \equiv 1+\sec A\,\mathrm{cosec}\,A$	1	220
0	0	2	0	2	0	$(\sec A-\tan A)(\sec A+\tan A) \equiv 1$	1	194
						$\dfrac{\sec^2 A - \tan^2 A + \tan A}{\sec A} \equiv \sin A + \cos A$	1	192
						$\dfrac{\tan A + \sec A - 1}{\tan A - \sec A + 1} \equiv \tan A + \sec A$	1	208
						$\dfrac{\sec A \sec B}{1 + \tan A \tan B} \equiv \sec(A-B)$	2	272
0	0	2	1	2	0	$\dfrac{(\sec A - \tan A)^2 + 1}{\mathrm{cosec}\,A(\sec A - \tan A)} \equiv 2\tan A$	1	249
0	0	3	0	0	0	$\tan A + \tan(A+120°) + \tan(A+240°)$ $\equiv 3\tan 3A$	3	391
						$\tan A + \tan B + \tan C \equiv \tan A \tan B \tan C$ (where $A+B+C=180°$)	3	356
						$\dfrac{3\tan A - \tan^3 A}{1 - 3\tan^2 A} \equiv \tan 3A$	2	325
						$\dfrac{\tan A}{\tan 2A - \tan A} \equiv \cos 2A$	2	302
0	0	4	0	0	0	$\dfrac{1 - \tan A \tan B}{1 + \tan A \tan B} \equiv \dfrac{\cos(A+B)}{\cos(A-B)}$	2	277
						$\dfrac{\tan A + \tan B}{\tan A - \tan B} \equiv \dfrac{\sin(A+B)}{\sin(A-B)}$	2	278

sin	cos	tan	cosec	sec	cot		Level	Page
0	0	4	0	0	0	$\dfrac{1+\tan A}{1-\tan A}+\dfrac{1-\tan A}{1+\tan A}\equiv 2\sec 2A$	2	332
0	0	4	0	0	4	$(\tan A+\tan B)(1-\cot A\cot B)$ $\qquad+(\cot A+\cot B)(1-\tan A\tan B)\equiv 0$	1	114
0	1	0	0	0	0	$2\cos^2(45°-A)\equiv 1+\sin 2A$	2	307
						$\dfrac{1}{8}(1-\cos 4A)\equiv \sin^2 A\cos^2 A$	2	314
						$1-2\cos^2 A\equiv 2\sin^2 A-1$	1	149
						$\dfrac{2}{1+\cos A}\equiv \sec^2\dfrac{A}{2}$	2	292
						$\dfrac{2}{1-\cos A}\equiv \mathrm{cosec}^2\dfrac{A}{2}$	2	291
0	1	0	0	0	2	$\dfrac{1-\cot^2 A}{1+\cot^2 A}+2\cos^2 A\equiv 1$	1	240
0	1	0	0	1	0	$\sec A-\cos A\equiv \sin A\tan A$	1	143
						$(1-\cos A)(1+\sec A)\equiv \sin A\tan A$	1	199
						$\dfrac{1+\sec A}{1+\cos A}\equiv \sec A$	1	160
0	1	0	0	1	1	$(\cos A+\cot A)\sec A\equiv 1+\mathrm{cosec}\,A$	1	141
0	1	0	1	0	0	$\cos A\,\mathrm{cosec}\,A\equiv \cot A$	1	83
						$\dfrac{\mathrm{cosec}\,A}{1-\cos A}\equiv \dfrac{1+\cos A}{\sin^3 A}$	1	226

sin	cos	tan	cosec	sec	cot		Level	Page
0	1	0	1	0	1	$\cos^2 A(\operatorname{cosec}^2 A - \cot^2 A) \equiv \cos^2 A$	1	150
						$\dfrac{\cot A \cos A}{\operatorname{cosec}^2 A - 1} \equiv \sin A$	1	123
0	1	1	0	0	0	$\cos A \tan A \equiv \sin A$	1	82
						$\cos^2 A(1 + \tan^2 A) \equiv 1$	1	117
						$\cos^2 A \tan A \equiv \sin A \cos A$	1	86
						$(1 + \cos A)\tan \dfrac{A}{2} \equiv \sin A$	2	293
						$(\cos^2 A - 1)(\tan^2 A + 1) \equiv -\tan^2 A$	1	122
						$\dfrac{1}{\cos^2 A(1 + \tan^2 A)} \equiv 1$	1	181
0	1	1	0	1	0	$\dfrac{\sec A - \cos A}{\tan A} \equiv \sin A$	1	207
0	1	1	0	1	1	$\dfrac{\sec A + \tan A}{\cot A + \cos A} \equiv \tan A \sec A$	1	210
0	1	1	1	0	0	$\cos A \tan A \operatorname{cosec} A \equiv 1$	1	85
						$\dfrac{\operatorname{cosec} A}{\cot A + \tan A} \equiv \cos A$	1	159
						$\dfrac{\cos A \operatorname{cosec} A}{\tan A} \equiv \cot^2 A$	1	126
0	2	0	0	0	0	$4\cos^3 A - 3\cos A \equiv \cos 3A$	2	340
						$\cos 3A + \cos 2A \equiv 2\cos \dfrac{5A}{2} \cos \dfrac{A}{2}$	3	358
						$8\cos^4 A - 4\cos 2A - 3 \equiv \cos 4A$	2	345

sin	cos	tan	cosec	sec	cot		Level	Page
0	2	0	0	0	0	$\dfrac{\cos A + 1}{\cos A - 1} \equiv \dfrac{1 + \sec A}{1 - \sec A}$	1	79
						$\dfrac{1}{1 - \cos A} + \dfrac{1}{1 + \cos A} \equiv 2\operatorname{cosec}^2 A$	1	176
						$\dfrac{1 - \cos A}{1 + \cos A} \equiv (\operatorname{cosec} A - \cot A)^2$	1	231
						$\cos(A+B)\cos(A-B) \equiv \cos^2 A - \sin^2 B$	2	308
						$\cos 4A + 4\cos 2A + 3 \equiv 8\cos^4 A$	2	342
						$\cos 2A - \cos 10A$ $\equiv \tan 4A(\sin 2A + \sin 10A)$	3	383
0	2	0	0	0	1	$\cos^2 A + \cot^2 A \cos^2 A \equiv \cot^2 A$	1	184
						$\cos A + \cos A \cot^2 A \equiv \cot A \operatorname{cosec} A$	1	147
0	2	0	0	1	0	$\cos A(\sec A - \cos A) \equiv \sin^2 A$	1	121
						$\dfrac{1 - \sec^2 A}{(1 - \cos A)(1 + \cos A)} \equiv -\sec^2 A$	1	179
0	2	0	0	2	0	$\dfrac{\sec A - \cos A}{\sec A + \cos A} \equiv \dfrac{\sin^2 A}{1 + \cos^2 A}$	1	190
0	2	0	2	0	0	$\dfrac{\cos A}{\operatorname{cosec} A - 1} + \dfrac{\cos A}{\operatorname{cosec} A + 1} \equiv 2\tan A$	1	205
0	2	1	0	0	0	$\cos^2 A + \tan^2 A \cos^2 A \equiv 1$	1	119
0	2	1	1	0	0	$\operatorname{cosec} A \tan \dfrac{A}{2} - \dfrac{\cos 2A}{1 + \cos A} \equiv 4\sin^2 \dfrac{A}{2}$	2	328
0	3	0	0	0	0	$32\cos^6 A - 48\cos^4 A + 18\cos^2 A - 1$ $\equiv \cos 6A$	2	341

sin	cos	tan	cosec	sec	cot		Level	Page
0	3	0	0	0	0	$\cos 3A + 2\cos 5A + \cos 7A$ $\equiv 4\cos^2 A \cos 5A$	3	385
						$\dfrac{\cos(A-B)}{\cos A \cos B} \equiv 1 + \tan A \tan B$	2	255
						$1 - \cos 2A + \cos 4A - \cos 6A$ $\equiv 4\sin A \cos 2A \sin 3A$	3	380
						$1 + \cos 2A + \cos 4A + \cos 6A$ $\equiv 4\cos A \cos 2A \cos 3A$	3	386
						$\dfrac{\cos(A+B)}{\cos A \cos B} \equiv 1 - \tan A \tan B$	2	275
						$\dfrac{\cos A + \cos 3A}{2\cos 2A} \equiv \cos A$	3	368
						$\cos A + \cos B + \cos C$ $\equiv 4\sin\dfrac{A}{2}\sin\dfrac{B}{2}\sin\dfrac{C}{2} + 1$ (where $A + B + C = 180°$)	3	357
0	4	0	0	0	0	$\dfrac{\cos 4A - \cos 8A}{\cos 4A + \cos 8A} \equiv \tan 2A \tan 6A$	3	377
						$\dfrac{\cos 2A - \cos 4A}{\cos 2A + \cos 4A} \equiv \tan 3A \tan A$	3	381
						$\dfrac{\cos A + \cos B}{\cos A - \cos B}$ $\equiv -\cot\left(\dfrac{A+B}{2}\right)\cot\left(\dfrac{A-B}{2}\right)$	3	362

sin	cos	tan	cosec	sec	cot		Level	Page
1	0	0	0	0	0	$1 - 2\sin^2 A \equiv 2\cos^2 A - 1$	1	155
						$\sin(A+B+C) \equiv \begin{cases} \sin A \cos B \cos C \\ +\sin B \cos C \cos A \\ +\sin C \cos A \cos B \\ -\sin A \sin B \sin C \end{cases}$	3	352
1	0	0	0	0	1	$\sin A \cot A \equiv \cos A$	1	80
1	0	0	0	1	0	$(1 - \sin^2 A)\sec^2 A \equiv 1$	1	88
						$\dfrac{\sec A}{1 - \sin A} \equiv \dfrac{1 + \sin A}{\cos^3 A}$	1	242
						$\sin A \sec A \equiv \tan A$	1	81
						$\dfrac{1 - \sin A}{\sec A} \equiv \dfrac{\cos^3 A}{1 + \sin A}$	1	237
1	0	0	1	0	0	$\operatorname{cosec} A - \sin A \equiv \cot A \cos A$	1	140
						$(1 - \sin A)(1 + \operatorname{cosec} A) \equiv \cos A \cot A$	1	200
						$\dfrac{1 + \sin A}{1 + \operatorname{cosec} A} \equiv \sin A$	1	75
1	0	1	0	0	0	$\dfrac{\tan A}{\sin A} \equiv \sec A$	1	105
						$\sin A \tan \dfrac{A}{2} \equiv 1 - \cos A$	2	297
						$\sin 2A \tan A \equiv 1 - \cos 2A$	2	298
						$\tan A \sin A \equiv \sec A - \cos A$	1	197

sin	cos	tan	cosec	sec	cot		Level	Page
1	0	1	0	0	0	$\tan^2 A - \sin^2 A \equiv \tan^2 A \sin^2 A$	1	151
1	0	1	0	1	0	$\sec A - \sin A \tan A \equiv \cos A$	1	139
						$(1 - \sin A)(\sec A + \tan A) \equiv \cos A$	1	196
						$\dfrac{\sin A \sec A}{\tan A} \equiv 1$	1	91
						$\dfrac{1 + \sec A}{\tan A + \sin A} \equiv \operatorname{cosec} A$	1	204
						$\dfrac{\tan A \sin A}{\sec^2 A - 1} \equiv \cos A$	1	127
1	1	0	0	0	0	$3 \sin^2 A + 4 \cos^2 A \equiv 3 + \cos^2 A$	1	111
						$\dfrac{\cos 2A}{1 + \sin 2A} \equiv \dfrac{\cot A - 1}{\cot A + 1}$	2	322
						$\dfrac{\sin 2A}{1 - \cos 2A} \equiv \cot A$	2	267
						$1 - \dfrac{\cos^2 A}{1 + \sin A} \equiv \sin A$	1	193
						$\cos^4 A - \sin^4 A \equiv \cos^2 A - \sin^2 A$	1	102
						$\cos^4 A - \sin^4 A \equiv \dfrac{1}{\sec 2A}$	2	253
						$\cos^4 A - \sin^4 A \equiv \cos 2A$	2	260
						$\dfrac{\sin A}{1 + \cos A} \equiv \tan \dfrac{A}{2}$	2	262
						$\dfrac{\sin 2A}{1 + \cos 2A} \equiv \tan A$	2	305
						$\dfrac{1 - \cos A}{\sin A} \equiv \tan \dfrac{A}{2}$	2	263

sin	cos	tan	cosec	sec	cot		Level	Page
1	1	0	0	0	0	$\cos^2 A - \sin^2 A \equiv 2\cos^2 A - 1$	1	118
						$1 - \dfrac{\sin^2 A}{1 - \cos A} \equiv -\cos A$	1	174
						$1 - \dfrac{\sin^2 A}{1 + \cos A} \equiv \cos A$	1	227
						$\sin^4 A + \cos^4 A \equiv \dfrac{3}{4} + \dfrac{1}{4}\cos 4A$	2	344
						$(\cos^2 A - 2)^2 - 4\sin^2 A \equiv \cos^4 A$	1	153
						$\sin(30° + A) + \cos(60° + A) \equiv \cos A$	2	265
						$\cos^2 A - \sin^2 A \equiv 1 - 2\sin^2 A$	1	87
						$\cos^2 A - \sin^2 A \equiv \cos^4 A - \sin^4 A$	1	102
						$1 - 8\sin^2 A \cos^2 A \equiv \cos 4A$	2	346
						$\left(\sin\dfrac{A}{2} + \cos\dfrac{A}{2}\right)^2 \equiv 1 + \sin A$	2	264
						$1 - \dfrac{\cos^2 A}{1 + \sin A} \equiv \sin A$	1	168
						$1 + \dfrac{\cos^2 A}{\sin A - 1} \equiv -\sin A$	1	232
						$\dfrac{\cos^2 A}{1 - \sin A} \equiv 1 + \sin A$	1	125
						$\dfrac{1 - \cos A}{\sin A} \equiv \dfrac{\sin A}{1 + \cos A}$	1	165
1	1	0	0	0	1	$\sin A + \cos A \cot A \equiv \operatorname{cosec} A$	1	138

sin	cos	tan	cosec	sec	cot		Level	Page
1	1	0	1	1	0	$\sin A \cos A \operatorname{cosec} A \sec A \equiv 1$	1	84
						$\dfrac{\sin A}{\operatorname{cosec} A} + \dfrac{\cos A}{\sec A} \equiv 1$	1	124
						$\dfrac{\sec A}{\operatorname{cosec} A} + \dfrac{\sin A}{\cos A} \equiv 2\tan A$	1	95
1	1	1	0	0	0	$\cos A + \sin A \tan A \equiv \sec A$	1	137
						$\tan A + \dfrac{\cos A}{1 + \sin A} \equiv \sec A$	1	223
						$\dfrac{\cos^2 A - \sin^2 A}{1 - \tan^2 A} \equiv \cos^2 A$	1	116
						$\dfrac{\tan A + \cos A}{\sin A} \equiv \sec A + \cot A$	1	92
1	1	1	0	0	1	$\tan^2 A \cos^2 A + \cot^2 A \sin^2 A \equiv 1$	1	112
						$\tan A(\sin A + \cot A \cos A) \equiv \sec A$	1	198
						$\dfrac{\sin^2 A - \tan A}{\cos^2 A - \cot A} \equiv \tan^2 A$	1	241
						$\dfrac{1 - \sin^2 A}{1 - \cos^2 A} + \tan A \cot A \equiv \operatorname{cosec}^2 A$	1	130
						$\dfrac{\cos A}{1 - \tan A} + \dfrac{\sin A}{1 - \cot A} \equiv \sin A + \cos A$	1	206
1	1	1	1	0	0	$\tan A \cos A + \operatorname{cosec} A \sin^2 A \equiv 2\sin A$	1	120
1	1	1	1	1	1	$(\operatorname{cosec} A - \sin A)(\sec A - \cos A)$ $\cdot (\tan A + \cot A) \equiv 1$	1	167

sin	cos	tan	cosec	sec	cot		Level	Page
1	2	0	0	0	0	$\dfrac{\sin^3 A + \cos^3 A}{1 - 2\cos^2 A} \equiv \dfrac{\sec A - \sin A}{\tan A - 1}$	1	191
						$\dfrac{2\sin(A - B)}{\cos(A + B) - \cos(A - B)} \equiv \cot A - \cot B$	2	321
						$\dfrac{(2\cos^2 A - 1)^2}{\cos^4 A - \sin^4 A} \equiv 1 - 2\sin^2 A$	1	246
						$\dfrac{\sin A + \cos A}{\cos A} \equiv 1 + \tan A$	1	76
						$\dfrac{\cos A}{\cos A - \sin A} \equiv \dfrac{1}{1 - \tan A}$	1	77
						$\dfrac{1 - 2\cos^2 A}{\sin A \cos A} \equiv \tan A - \cot A$	1	175
						$\dfrac{\cos(A + B)}{\sin A \cos A} \equiv \cot A - \cot B$	2	256
						$\dfrac{\cos(A - B)}{\sin A \cos B} \equiv \cot A + \tan B$	2	257
						$\dfrac{\sin(A + B)}{\cos A \cos B} \equiv \tan A + \tan B$	2	259
						$\dfrac{\cos(A + B)}{\cos A \sin B} \equiv \cot B - \tan A$	2	271
						$\dfrac{5 - 10\cos^2 A}{\sin A - \cos A} \equiv 5(\sin A + \cos A)$	1	228
1	3	0	0	0	0	$\dfrac{\cos A}{1 + \cos 2A} + \dfrac{\sin A}{1 - \cos 2A} \equiv \dfrac{\sin A + \cos A}{\sin 2A}$	2	331

sin	cos	tan	cosec	sec	cot		Level	Page
1	3	0	0	0	0	$\dfrac{\sin A\left(4\cos^2 A-1\right)}{\cos A\cos 2A}\equiv \tan A+\tan 2A$	2	337
2	0	0	0	0	0	$\sin 5A-\sin 3A\equiv 2\sin A\cos 4A$	3	359
						$3\sin A-4\sin^3 A\equiv \sin 3A$	2	324
						$2\sin^2\dfrac{A}{6}-\sin^2\dfrac{A}{7}\equiv \cos^2\dfrac{A}{7}-\cos\dfrac{A}{3}$	2	304
						$\dfrac{\sin^2 A}{1-\sin^2 A}\equiv \tan^2 A$	1	100
						$\dfrac{1+\sin A}{1-\sin A}\equiv (\sec A+\tan A)^2$	1	229
						$\dfrac{1+\sin A}{1-\sin A}\equiv \dfrac{\operatorname{cosec} A+1}{\operatorname{cosec} A-1}$	1	78
						$\dfrac{1-\sin A}{1+\sin A}\equiv (\sec A-\tan A)^2$	1	248
						$\sin(A+B)\sin(A-B)\equiv \cos^2 B-\cos^2 A$	2	311
						$\dfrac{1}{\sin A+1}-\dfrac{1}{\sin A-1}\equiv 2\sec^2 A$	1	177
						$2\sin 2A(1-2\sin^2 A)\equiv \sin 4A$	2	339
						$\sin(A+B)\sin(A-B)\equiv \sin^2 A-\sin^2 B$	2	310
2	0	0	0	0	1	$\sin A+\sin A\cot^2 A\equiv \operatorname{cosec} A$	1	195
2	0	0	1	0	0	$\sin A(\operatorname{cosec} A-\sin A)\equiv \cos^2 A$	1	134

sin	cos	tan	cosec	sec	cot		Level	Page
2	0	1	0	0	0	$\tan 4A(\sin 2A + \sin 10A)$ $\equiv \cos 2A - \cos 10A$	3	388
						$\dfrac{\sin A + \tan A}{\sin A} \equiv 1 + \sec A$	1	94
						$\sin A + \sin A \tan^2 A \equiv \tan A \sec A$	1	154
2	1	0	0	0	0	$(4 \sin A \cos A)(1 - 2 \sin^2 A) \equiv \sin 4A$	2	283
						$\dfrac{\left(2 \sin^2 A - 1\right)^2}{\sin^4 A - \cos^4 A} \equiv 1 - 2 \cos^2 A$	1	136
						$\dfrac{\sin^2 A + \cos^2 A}{\sin^2 A} \equiv \operatorname{cosec}^2 A$	1	99
						$\dfrac{\sin A}{\sin A - \cos A} \equiv \dfrac{1}{1 - \cot A}$	1	93
						$\dfrac{1 - 2 \sin^2 A}{\sin A \cos A} \equiv \cot A - \tan A$	1	169
						$\dfrac{1 + \sin A - \sin^2 A}{\cos A} \equiv \cos A + \tan A$	1	96
						$\dfrac{\cos(A - B)}{\sin A \sin B} \equiv \cot A \cot B + 1$	1	258
						$\dfrac{\sin(A + B)}{\sin A \cos B} \equiv \cot A \tan B + 1$	2	276
						$\dfrac{\sin(A - B)}{\sin A \cos B} \equiv 1 - \cot A \tan B$	2	254
2	1	0	0	1	0	$\dfrac{\sec A}{\sin A} - \dfrac{\sin A}{\cos A} \equiv \cot A$	1	213

sin	cos	tan	cosec	sec	cot		Level	Page
2	1	0	0	2	0	$2\sec^2 A - 2\sec^2 A \sin^2 A$ $- \sin^2 A - \cos^2 A \equiv 1$	1	187
2	2	0	0	0	0	$(\cos A - \sin A)^2 + 2\sin A \cos A \equiv 1$	1	133
						$(a\sin A + b\cos A)^2 + (a\cos A - b\sin A)^2$ $\equiv a^2 + b^2$	1	106
						$(2a\sin A \cos A)^2 + a^2(\cos^2 A - \sin^2 A)^2 \equiv a^2$	2	261
						$\dfrac{\sin 3A}{\sin A} - \dfrac{\cos 3A}{\cos A} \equiv 2$	2	287
						$\dfrac{1 + \sin A}{\cos A} + \dfrac{\cos A}{1 + \sin A} \equiv \dfrac{2}{\cos A}$	1	164
						$\dfrac{\cos A}{1 - \sin^2 A - \cos^2 A + \sin A} \equiv \cot A$	1	129
						$\dfrac{\sin A \cos A}{\cos^2 A - \sin^2 A} \equiv \dfrac{\tan A}{1 - \tan^2 A}$	1	98
						$\dfrac{\sin(A + 45°)}{\cos(A + 45°)} + \dfrac{\cos(A + 45°)}{\sin(A + 45°)} \equiv 2\sec 2A$	2	336
						$\dfrac{\cos 2A}{\sin A} + \dfrac{\sin 2A}{\cos A} \equiv \operatorname{cosec} A$	2	320
						$\dfrac{\sin^2 2A + 2\cos 2A - 1}{\sin^2 2A + 3\cos 2A - 3} \equiv \dfrac{1}{1 - \sec 2A}$	2	335
						$\dfrac{\sin A}{1 - \cos A} + \dfrac{1 - \cos A}{\sin A} \equiv 2\operatorname{cosec} A$	1	221
						$\dfrac{\cos A}{1 + \sin A} + \dfrac{1 + \sin A}{\cos A} \equiv 2\sec A$	1	216
						$\dfrac{1 - \sin A}{\cos A} + \dfrac{\cos A}{1 - \sin A} \equiv 2\sec A$	1	214

sin	cos	tan	cosec	sec	cot		Level	Page
2	2	0	0	0	0	$\dfrac{1+\cos A}{\sin A}+\dfrac{\sin A}{1+\cos A}\equiv 2\operatorname{cosec} A$	1	212
						$\dfrac{\sin A}{1+\cos A}+\dfrac{1+\cos A}{\sin A}\equiv\dfrac{2}{\sin A}$	1	161
						$\dfrac{1-\cos 2A+\sin A}{\sin 2A+\cos A}\equiv \tan A$	2	347
						$\dfrac{\sin A+\sin 2A}{2+3\cos A+\cos 2A}\equiv \tan\dfrac{A}{2}$	2	327
						$\dfrac{\sin A+\sin 2A}{1+\cos A+\cos 2A}\equiv \tan A$	2	268
						$(\sin A+\cos A)^2+(\sin A-\cos A)^2\equiv 2$	1	156
						$\dfrac{\sin(2A+B)+\sin B}{\cos(2A+B)+\cos B}\equiv \tan(A+B)$	3	361
						$\dfrac{\sin(A+B)-\sin(A-B)}{\cos(A+B)+\cos(A-B)}\equiv \tan B$	3	370
						$\dfrac{\sin 4A-\sin 2A}{\cos 4A+\cos 2A}\equiv \tan A$	3	366
						$\dfrac{\sin 2A+\cos 2A+1}{\sin 2A+\cos 2A-1}\equiv\dfrac{\tan(45°+A)}{\tan A}$	2	333
						$\dfrac{1+\sin 2A+\cos 2A}{\sin A+\cos A}\equiv 2\cos A$	2	266
						$\dfrac{\cos^2 A}{\sin^2 A}+\cos^2 A+\sin^2 A\equiv\dfrac{1}{\sin^2 A}$	1	132

sin	cos	tan	cosec	sec	cot		Level	Page
2	2	0	0	0	0	$\dfrac{1+\sin A+\cos A}{1+\sin A-\cos A} \equiv \dfrac{1+\cos A}{\sin A}$	1	211
						$\dfrac{1+\cos A+\sin A}{1+\cos A-\sin A} \equiv \sec A+\tan A$	1	217
						$\dfrac{\sin^3 A+\cos^3 A}{\sin A+\cos A} \equiv 1-\sin A\cos A$	1	107
						$\dfrac{\cos^3 A-\sin^3 A}{\cos A-\sin A} \equiv \dfrac{2+\sin 2A}{2}$	2	286
						$\dfrac{\sin A-\cos A+1}{\sin A+\cos A-1} \equiv \dfrac{\sin A+1}{\cos A}$	1	209
						$\dfrac{\sin^3 A+\cos^3 A}{\sin A+\cos A} \equiv 1-\dfrac{1}{2}\sin 2A$	2	294
						$\dfrac{\sin 4A+\sin 2A}{\cos 4A+\cos 2A} \equiv \tan 3A$	3	373
						$\dfrac{\cos A-\cos 3A}{\sin 3A-\sin A} \equiv \tan 2A$	3	374
						$\dfrac{\cos A-\cos 3A}{\sin 3A+\sin A} \equiv \tan A$	3	367
						$\dfrac{\cos A-\cos 5A}{\sin A+\sin 5A} \equiv \tan 2A$	3	371
						$\dfrac{\sin 4A+\sin 8A}{\cos 4A+\cos 8A} \equiv \tan 6A$	3	375
						$\dfrac{\sin 4A-\sin 8A}{\cos 4A-\cos 8A} \equiv -\cot 6A$	3	376

sin	cos	tan	cosec	sec	cot		Level	Page
2	2	0	0	0	0	$\dfrac{\sin A + \sin B}{\cos A + \cos B} \equiv \tan\left(\dfrac{A+B}{2}\right)$	3	363
						$\dfrac{\sin A - \sin B}{\cos A - \cos B} \equiv -\cot\left(\dfrac{A+B}{2}\right)$	3	364
2	2	0	0	0	2	$\dfrac{\sin A \cos B}{\cos A \sin B}(\cot A \cot B) + 1 \equiv \dfrac{1}{\sin^2 B}$	1	234
3	0	0	0	0	0	$\sin 5A + 2\sin 3A + \sin A \equiv 4\sin 3A \cos^2 A$	3	379
						$\sin 2A + \sin 4A - \sin 6A$ $\equiv 4\sin A \sin 2A \sin 3A$	3	384
						$\sin 2A + \sin 2B + \sin 2C$ $\equiv 4\sin A \sin B \sin C$ (where $A+B+C=180°$)	3	354
						$\sin A + \sin B + \sin C$ $\equiv 4\cos\dfrac{A}{2}\cos\dfrac{B}{2}\cos\dfrac{C}{2}$ (where $A+B+C=180°$)	3	350
						$\dfrac{\sin A + \sin 3A}{2\sin 2A} \equiv \cos A$	3	365
						$\sin A(\sin 3A + \sin 5A)$ $\equiv \cos A(\cos 3A - \cos 5A)$	3	389
						$\sin A(\sin A + \sin 3A)$ $\equiv \cos A(\cos A - \cos 3A)$	3	390

sin	cos	tan	cosec	sec	cot		Level	Page
3	1	0	0	0	0	$\dfrac{\cos A + \sin A - \sin^3 A}{\sin A}$		
						$\equiv \cot A + \cos^2 A$	1	115
						$\dfrac{\sin A - \sin 3A}{\sin^2 A - \cos^2 A} \equiv 2\sin A$	3	369
3	1	0	2	0	1	$\operatorname{cosec} A(\operatorname{cosec} A - \sin A)$		
						$+\left(\dfrac{\sin A - \cos A}{\sin A}\right) + \cot A \equiv \operatorname{cosec}^2 A$	1	235
3	2	0	0	0	0	$\dfrac{\sin 3A \cos A - \sin A \cos 3A}{\sin 2A} \equiv 1$	2	284
3	3	0	0	0	0	$(\sin A + \cos B)^2$		
						$+ (\cos B + \sin A)(\cos B - \sin A)$		
						$\equiv 2\cos B(\sin A + \cos B)$	1	108
						$(\sin A - \cos B)^2$		
						$+ (\cos B + \sin A)(\cos B - \sin A)$		
						$\equiv -2\cos B(\sin A - \cos B)$	1	202
						$\dfrac{\sin A + \cos A}{\cos A} - \dfrac{\sin A - \cos A}{\sin A}$		
						$\equiv \sec A \operatorname{cosec} A$	1	173
						$\dfrac{\sin A + \sin 2A + \sin 3A}{\cos A + \cos 2A + \cos 3A} \equiv \tan 2A$	3	372

sin	cos	tan	cosec	sec	cot		Level	Page
3	3	0	0	0	0	$\dfrac{\sin A + \cos A}{\sin A} - \dfrac{\cos A - \sin A}{\cos A}$ $\equiv \operatorname{cosec} A \sec A$	1	170
3	6	0	0	0	0	$\sin A \cos B \cos C + \sin B \cos A \cos C$ $\quad + \sin C \cos A \cos B$ $\equiv \sin A \sin B \sin C$ $\quad (A + B + C = 180°)$	3	353
4	0	0	0	0	0	$\dfrac{\sin A + \sin B}{\sin A - \sin B}$ $\equiv \tan\left(\dfrac{A+B}{2}\right)\cot\left(\dfrac{A-B}{2}\right)$	3	360
						$\dfrac{\sin 4A + \sin 8A}{\sin 4A - \sin 8A} \equiv -\dfrac{\tan 6A}{\tan 2A}$	3	378
						$\dfrac{1 + \sin A}{1 - \sin A} - \dfrac{1 - \sin A}{1 + \sin A} \equiv 4 \tan A \sec A$	1	172
4	0	2	0	0	0	$\dfrac{\sin 2A + \sin 4A}{\sin 2A - \sin 4A} + \dfrac{\tan 3A}{\tan A} \equiv 0$	3	382
4	2	0	0	0	0	$\dfrac{\sin 2A \cos A - 2\cos 2A \sin A}{2 \sin A - \sin 2A} \equiv 2 \cos^2 \dfrac{A}{2}$	2	330
4	4	0	0	0	0	$\dfrac{\cos A - \cos B}{\sin A + \sin B} + \dfrac{\sin A - \sin B}{\cos A + \cos B} \equiv 0$	2	281
						$\dfrac{\cos A - \sin A}{\cos A + \sin A} + \dfrac{\cos A + \sin A}{\cos A - \sin A} \equiv 2 \sec 2A$	2	279
						$\dfrac{\cos A + \sin A}{\cos A - \sin A} - \dfrac{\cos A - \sin A}{\cos A + \sin A} \equiv 2 \tan 2A$	2	282

The Encyclopedia of
Trigonometric Games or
Trigonometric Identity Proofs
(TIPs)

Level-One-Games

Easy Proofs

An Example for Proving a Trig Identity

$$\frac{1 + \sin A}{1 + \operatorname{cosec} A} \equiv \sin A$$

Eyeballing and Mental Gymnastics

1. *Start with the more complex side, normally the LHS; this is more amenable to simplification.*
2. *Consider simplifying tan, cot, and the reciprocal functions to sin and cos.*
3. *Consider the use of common denominators.*
4. *Rearrange and simplify through cancellation of common terms, if available.*

Let us explore this first game (proof) together. Eyeballing the identity, we see that the more complex side is indeed the LHS (occassionally the RHS is the more complex; then it may be preferable to begin with the RHS. On rare occassions both the LHS and the RHS are complex; then one can explore simplifying both side to a common set of terms).

We note the cosec term in the denominator and remember that cosec = 1/ sin.

Generally speaking it is easier to work with sin and cos than with their reciprocals as it makes rearrangement, simplification and cancellation easier.

With a reciprocal term in the denominator, we expect to use common denominators prior to rearrangement, simplification and cancellation.

If all goes well, *and no careless mistakes are made*, we should end up with "sin *A*" which is the target objective on the RHS.

It is useful to begin a proof by writing down *accurately* the LHS. If a mistake is made here, no amount of effort will give the required identity.

$$\boxed{\frac{1+\sin A}{1+\operatorname{cosec} A} \equiv \sin A}$$

$$\text{LHS} = \frac{1+\sin A}{1+\operatorname{cosec} A}$$

$$= \frac{1+\sin A}{\left(1+\dfrac{1}{\sin A}\right)}$$

$$= (1+\sin A)\left(\frac{1}{\dfrac{\sin A + 1}{\sin A}}\right)$$

$$= (1+\sin A)\left(\frac{\sin A}{\sin A + 1}\right)$$

$$= \sin A$$

$$\equiv \text{RHS}.$$

> *Use a bracket, if it helps focus on the key groups and minimises careless mistakes.*

$$\boxed{\frac{\sin A + \cos A}{\cos A} \equiv 1 + \tan A}$$

Eyeballing and Mental Gymnastics*

1. $t = s/c$
2. *rearrange and simplify.*

$$\text{LHS} = \frac{\sin A + \cos A}{\cos A}$$

$$= \tan A + 1$$

$$\equiv \text{RHS}.$$

---

*Some teachers prefer that students always write out the trig functions to include the angle i.e. $\sin A$ instead of sin. In the "Eyeballing and Mental Gymnastics" sections, and in the short explanatory notes, we will use the abbreviations: $s = \sin$, $c = \cos$, and $t = \tan$. Such abbreviations reflect the mental process in action, and conveys a sense of speed with eyeballing and mental gymnastics taking place.

$$\boxed{\frac{\cos A}{\cos A - \sin A} \equiv \frac{1}{1 - \tan A}}$$

Eyeballing and Mental Gymnastics

1. *divide* LHS *by* cos
2. $t = s/c$

$$\text{LHS} = \frac{\cos A}{\cos A - \sin A}$$

$$= \frac{1}{1 - \tan A}$$

$$\equiv \text{RHS.}$$

$$\boxed{\frac{1+\sin A}{1-\sin A} \equiv \frac{\csc A + 1}{\csc A - 1}}$$

Eyeballing and Mental Gymnastics

1. $\csc = 1/\sin$
2. *rearrange and simplify.*

$$\text{LHS} = \frac{1+\sin A}{1-\sin A}$$

$$= \frac{\csc + 1}{\csc - 1}$$

divide both
numerator
and denominator
by sin

$$\equiv \text{RHS}.$$

$$\boxed{\frac{\cos A + 1}{\cos A - 1} \equiv \frac{1 + \sec A}{1 - \sec A}}$$

Eyeballing and Mental Gymnastics

1. $\sec = 1/\cos$
2. *rearrange and simplify.*

$$\text{LHS} = \frac{\cos A + 1}{\cos A - 1}$$

$$= \frac{1 + \sec A}{1 - \sec A}$$

$$\equiv \text{RHS}.$$

divide both
numerator
and denominator
by cos

$$\boxed{\sin A \cot A \equiv \cos A}$$

Eyeballing and Mental Gymnastics

1. $\cot = c/s$
2. *simplify.*

$$
\begin{aligned}
\text{LHS} &= \sin A \cot A \\
&= \sin A \cdot \frac{\cos A}{\sin A} \\
&= \cos A \\
&\equiv \text{RHS.}
\end{aligned}
$$

$$\boxed{\sin A \sec A \equiv \tan A}$$

Eyeballing and Mental Gymnastics

1. $\sec = 1/\cos$, $t = s/c$
2. *simplify.*

$$
\begin{aligned}
\text{LHS} &= \sin A \cdot \sec A \\
&= \sin A \cdot \frac{1}{\cos A} \\
&= \tan A \\
&\equiv \text{RHS.}
\end{aligned}
$$

$$\boxed{\cos A \tan A \equiv \sin A}$$

Eyeballing and Mental Gymnastics

1. $t = s/c$
2. *simplify.*

$$\text{LHS} = \cos A \tan A$$

$$= \cos A \frac{\sin A}{\cos A}$$

$$= \sin A$$

$$\equiv \text{RHS}.$$

$$\boxed{\cos A \operatorname{cosec} A \equiv \cot A}$$

Eyeballing and Mental Gymnastics

1. $\operatorname{cosec} = 1/\sin$, $\cot = c/s$
2. *simplify.*

$$\begin{aligned}
\text{LHS} &= \cos A \operatorname{cosec} A \\
&= \cos A \cdot \frac{1}{\sin A} \\
&= \cot A \\
&\equiv \text{RHS.}
\end{aligned}$$

$$\sin A \cos A \operatorname{cosec} A \sec A \equiv 1$$

Eyeballing and Mental Gymnastics

1. $\operatorname{cosec} = 1/\sin$, $\sec = 1/\cos$
2. *simplify.*

$$\text{LHS} = \sin A \cos A \operatorname{cosec} A \sec A$$
$$= \sin A \cos A \cdot \frac{1}{\sin A} \cdot \frac{1}{\cos A}$$
$$= 1$$
$$\equiv \text{RHS.}$$

$$\boxed{\cos A \tan A \operatorname{cosec} A \equiv 1}$$

Eyeballing and Mental Gymnastics

1. $t = s/c$, $\operatorname{cosec} = 1/\sin$
2. *simplify.*

$$\begin{aligned}
\text{LHS} &= \cos A \tan A \operatorname{cosec} A \\
&= \cos A \cdot \frac{\sin A}{\cos A} \cdot \frac{1}{\sin A} \\
&= 1 \\
&\equiv \text{RHS.}
\end{aligned}$$

$$\boxed{\tan A \cos^2 A \equiv \sin A \cos A}$$

Eyeballing and Mental Gymnastics

1. $t = s/c$
2. *simplify.*

$$\text{LHS} = \tan A \cos^2 A$$

$$= \frac{\sin A}{\cos A} \cdot \cos^2 A$$

$$= \sin A \cos A$$

$$\equiv \text{RHS.}$$

$$\boxed{\cos^2 A - \sin^2 A \equiv 1 - 2\sin^2 A}$$

Eyeballing and Mental Gymnastics

1. c^2, s^2 suggest $s^2 + c^2 \equiv 1$*
2. *simplify.*

$$\text{LHS} = \cos^2 A - \sin^2 A$$

$$= (1 - \sin^2 A) - \sin^2 A$$

$$= 1 - 2\sin^2 A$$

$$\equiv \text{RHS}.$$

---

*We shall use the abbreviation: $s^2 + c^2 \equiv 1$ for the trig equivalent of Pythagoras' Theorem $\sin^2 A + \cos^2 A \equiv 1$.

$$\boxed{\left(1 - \sin^2 A\right) \sec^2 A \equiv 1}$$

Eyeballing and Mental Gymnastics

1. s^2 *suggests* $s^2 + c^2 \equiv 1$
2. $\sec = 1/\cos$
3. *simplify.*

$$\begin{aligned}
\text{LHS} &= \left(1 - \sin^2 A\right) \sec^2 A \\
&= \cos^2 A \cdot \frac{1}{\cos^2 A} \\
&= 1 \\
&\equiv \text{RHS.}
\end{aligned}$$

$$\boxed{\tan A(\cot A + \tan A) \equiv \sec^2 A}$$

Eyeballing and Mental Gymnastics

1. $t = s/c$, $\cot = c/s$
2. $\sec^2$ *suggests* $s^2 + c^2 = 1^*$
3. *simplify.*

$$\text{LHS} = \tan A(\cot A + \tan A)$$

$$= 1 + \tan^2 A$$

$$= \sec^2 A$$

$$\equiv \text{RHS}.$$

---

*The identity $s^2 + c^2 \equiv 1$ should also trigger off possible reference to:

$$\tan^2 A + 1 \equiv \sec^2 A \ (\text{divide } s^2 + c^2 \equiv 1 \text{ by } c^2)$$

and

$$1 + \cot^2 A \equiv \text{cosec}^2 A \ (\text{divide } s^2 + c^2 \equiv 1 \text{ by } s^2).$$

$$\boxed{\tan^2 A(\operatorname{cosec}^2 A - 1) \equiv 1}$$

Eyeballing and Mental Gymnastics

1. $\tan^2$, cosec^2 *suggest* $s^2 + c^2 \equiv 1$
2. *rearrange and simplify.*

$$\begin{aligned}
\text{LHS} &= \tan^2 A(\operatorname{cosec}^2 A - 1) \\
&= \tan^2 A(\cot^2 A) \\
&\equiv 1 \\
&\equiv \text{RHS.}
\end{aligned}$$

$$\boxed{\frac{\sin A \sec A}{\tan A} \equiv 1}$$

Eyeballing and Mental Gymnastics

1. $\sec = 1/\cos$, $t = s/c$
2. *simplify.*

$$\text{LHS} = \frac{\sin A \sec A}{\tan A}$$

$$= \sin A \cdot \frac{1}{\cos A} \cdot \frac{\cos A}{\sin A}$$

$$\equiv 1$$

$$\equiv \text{RHS}.$$

$$\boxed{\frac{\tan A + \cos A}{\sin A} \equiv \sec A + \cot A}$$

Eyeballing and Mental Gymnastics

1. $t = s/c$, $\sec = 1/\cos$, $\cot = c/s$
2. *simplify.*

$$\text{LHS} = \frac{\tan A + \cos A}{\sin A}$$

$$= \frac{\sin A}{\cos A} \cdot \frac{1}{\sin A} + \frac{\cos A}{\sin A}$$

$$= \sec A + \cot A$$

$$\equiv \text{RHS.}$$

$$\frac{\sin A}{\sin A - \cos A} \equiv \frac{1}{1 - \cot A}$$

Eyeballing and Mental Gymnastics

1. $\cot = c/s$
2. *rearrange and simplify.*

$$\text{LHS} = \frac{\sin A}{\sin A - \cos A}$$

$$= \frac{1}{1 - \dfrac{\cos A}{\sin A}} \qquad \begin{vmatrix} \text{divide both} \\ \text{numerator} \\ \text{and denominator} \\ \text{by } \sin A \end{vmatrix}$$

$$= \frac{1}{1 - \cot A}$$

$$\equiv \text{RHS.}$$

$$\boxed{\frac{\sin A + \tan A}{\sin A} \equiv 1 + \sec A}$$

Eyeballing and Mental Gymnastics

1. $t = s/c$, $\sec = 1/\cos$
2. *simplify.*

$$\text{LHS} = \frac{\sin A + \tan A}{\sin A}$$

$$= \frac{\sin A}{\sin A} + \frac{\sin A}{\cos A} \cdot \frac{1}{\sin A}$$

$$= 1 + \sec A$$

$$\equiv \text{RHS.}$$

$$\frac{\sec A}{\operatorname{cosec} A} + \frac{\sin A}{\cos A} \equiv 2\tan A$$

Eyeballing and Mental Gymnastics

1. $\sec = 1/\cos$, $\operatorname{cosec} = 1/\sin$, $t = s/c$
2. *rearrange and simplify.*

$$\begin{aligned}
\text{LHS} &= \frac{\sec A}{\operatorname{cosec} A} + \frac{\sin A}{\cos A} \\
&= \frac{1}{\cos A} \cdot \frac{\sin A}{1} + \frac{\sin A}{\cos A} \\
&= \tan A + \tan A \\
&= 2\tan A.
\end{aligned}$$

$$\frac{1 + \sin A - \sin^2 A}{\cos A} \equiv \cos A + \tan A$$

Eyeballing and Mental Gymnastics

1. *s^2 suggests $s^2 + c^2 \equiv 1$*
2. *$t = s/c$*
3. *rearrange and simplify.*

$$\begin{aligned}
\text{LHS} &= \frac{1 + \sin A - \sin^2 A}{\cos A} \\
&= \frac{\cos^2 A + \sin A}{\cos A} \\
&= \cos A + \tan A \\
&= \text{RHS.}
\end{aligned}$$

$$\boxed{\frac{1}{\sec^2 A} + \frac{1}{\operatorname{cosec}^2 A} \equiv 1}$$

Eyeballing and Mental Gymnastics

1. $\sec = 1/\cos$, $\operatorname{cosec} = 1/\sin$
2. $\sec^2$, cosec^2 *suggest* $s^2 + c^2 \equiv 1$
3. *simplify.*

$$\text{LHS} = \frac{1}{\sec^2 A} + \frac{1}{\operatorname{cosec}^2 A}$$

$$= \cos^2 A + \sin^2 A$$

$$= 1$$

$$\equiv \text{RHS}.$$

$$\boxed{\frac{\sin A \cos A}{\cos^2 A - \sin^2 A} \equiv \frac{\tan A}{1 - \tan^2 A}}$$

Eyeballing and Mental Gymnastics

1. $t = s/c$
2. *divide both numerator and denominator by* $\cos^2 A$ *to give* $\cos^2 A / \cos^2 A = 1$
3. *rearrange and simplify.*

$$\text{LHS} = \frac{\sin A \cos A}{\cos^2 A - \sin^2 A}$$

$$= \frac{\dfrac{\sin A \cos A}{\cos^2 A}}{\dfrac{\cos^2 A - \sin^2 A}{\cos^2 A}}$$

$$= \frac{\tan A}{1 - \tan^2 A}$$

$$\equiv \text{RHS}.$$

$$\boxed{\frac{\sin^2 A + \cos^2 A}{\sin^2 A} \equiv \operatorname{cosec}^2 A}$$

Eyeballing and Mental Gymnastics

1. s^2, c^2 *suggest* $s^2 + c^2 \equiv 1$
2. $\operatorname{cosec} = 1/\sin$
3. *simplify.*

$$\text{LHS} = \frac{\sin^2 A + \cos^2 A}{\sin^2 A}$$

$$= \frac{1}{\sin^2 A}$$

$$= \operatorname{cosec}^2 A$$

$$\equiv \text{RHS.}$$

$$\boxed{\frac{\sin^2 A}{1 - \sin^2 A} \equiv \tan^2 A}$$

Eyeballing and Mental Gymnastics

1. s^2, t^2 suggest $s^2 + c^2 \equiv 1$
2. $t = s/c$.

$$\text{LHS} = \frac{\sin^2 A}{1 - \sin^2 A}$$

$$= \frac{\sin^2 A}{\cos^2 A}$$

$$= \tan^2 A$$

$$\equiv \text{RHS}.$$

$$\tan^2 A + \cot^2 A + 2 \equiv \operatorname{cosec}^2 A + \sec^2 A$$

Eyeballing and Mental Gymnastics

1. $\tan^2$, $\cot^2$, cosec^2, $\sec^2$ *suggest* $s^2 + c^2 \equiv 1$
2. *rearrange and simplify.*

$$
\begin{aligned}
\text{LHS} &= \tan^2 A + \cot^2 A + 2 \\
&= (\tan^2 A + 1) + (\cot^2 A + 1) \\
&= \sec^2 A + \operatorname{cosec}^2 A \\
&\equiv \text{RHS.}
\end{aligned}
$$

$$\boxed{\cos^4 A - \sin^4 A \equiv \cos^2 A - \sin^2 A}$$

Eyeballing and Mental Gymnastics

1. $\cos^4 - \sin^4$ *suggest* $a^4 - b^4 = (a^2 - b^2)(a^2 + b^2)$
2. c^2, s^2 *suggest* $s^2 + c^2 \equiv 1$
3. *rearrange and simplify.*

$$\text{LHS} = \cos^4 A - \sin^4 A$$

$$= (\cos^2 A - \sin^2 A)(\cos^2 A + \sin^2 A)$$

$$= (\cos^2 A - \sin^2 A)(1)$$

$$= \cos^2 A - \sin^2 A$$

$$\equiv \text{RHS.}$$

$$\boxed{\operatorname{cosec}^4 A - \cot^4 A \equiv \operatorname{cosec}^2 A + \cot^2 A}$$

Eyeballing and Mental Gymnastics

1. $(\operatorname{cosec}^4 - \cot^4)$ *suggest* $a^4 - b^4 = (a^2 - b^2)(a^2 + b^2)$
2. *squares suggest* $s^2 + c^2 \equiv 1$
3. *rearrange and simplify.*

$$\begin{aligned}
\text{LHS} &= \operatorname{cosec}^4 A - \cot^4 A \\
&= (\operatorname{cosec}^2 A - \cot^2 A)(\operatorname{cosec}^2 A + \cot^2 A) \\
&= (1)(\operatorname{cosec}^2 A + \cot^2 A) \\
&\equiv \text{RHS.}
\end{aligned}$$

$$\boxed{\tan A \, \csc A \equiv \sec A}$$

Eyeballing and Mental Gymnastics

1. $t = s/c$, $\csc = 1/\sin$, $\sec = 1/\cos$
2. *simplify.*

$$\text{LHS} = \tan A \, \csc A$$

$$= \frac{\sin A}{\cos A} \cdot \frac{1}{\sin A}$$

$$= \frac{1}{\cos A}$$

$$= \sec A$$

$$\equiv \text{RHS.}$$

$$\boxed{\frac{\tan A}{\sin A} \equiv \sec A}$$

Eyeballing and Mental Gymnastics

1. $t = s/c$
2. *simplify.*

$$\text{LHS} = \frac{\tan A}{\sin A}$$

$$= \frac{\sin A}{\cos A} \cdot \frac{1}{\sin A}$$

$$= \frac{1}{\cos A}$$

$$= \sec A$$

$$\equiv \text{RHS.}$$

$$(a \sin A + b \cos A)^2 + (a \cos A - b \sin A)^2 \equiv a^2 + b^2$$

Eyeballing and Mental Gymnastics

1. $(\ )^2$ *suggests expansion*
2. $s^2 + c^2 \equiv 1$
3. *rearrange and simplify.*

$$\text{LHS} = (a \sin A + b \cos A)^2 + (a \cos A - b \sin A)^2$$

$$= a^2 \sin^2 A + 2ab \sin A \cos A + b^2 \cos^2 A$$

$$+ a^2 \cos^2 A - 2ab \sin A \cos A + b^2 \sin^2 A$$

$$= a^2 (\sin^2 A + \cos^2 A) + b^2 (\sin^2 A + \cos^2 A)$$

$$= a^2 + b^2$$

$$\equiv \text{RHS.}$$

$$\boxed{\frac{\sin^3 A + \cos^3 A}{\sin A + \cos A} \equiv 1 - \sin A \cos A}$$

Eyeballing and Mental Gymnastics

1. $s^3 + c^3$ suggests $(s+c)(s^2 - cs + c^2)$
2. $s^2 + c^2 \equiv 1$
3. *rearrange and simplify.*

$$\text{LHS} = \frac{\sin^3 A + \cos^3 A}{\sin A + \cos A}$$

$$= \frac{(\sin A + \cos A)(\sin^2 A - \sin A \cos A + \cos^2 A)}{(\sin A + \cos A)}$$

$$= \sin^2 A - \sin A \cos A + \cos^2 A$$

$$= 1 - \sin A \cos A$$

$$\equiv \text{RHS}.$$

$$(\sin A + \cos B)^2 + (\cos B + \sin A)(\cos B - \sin A) \equiv 2\cos B(\sin A + \cos B)$$

Eyeballing and Mental Gymnastics

1. *Although the identity looks lengthy and complicated, closer inspection shows that the term* $(\sin A + \cos B)$ *is common to both sides of the equation.*
2. *rearrange and simplify.*

$$\begin{aligned}
\text{LHS} &= (\sin A + \cos B)^2 + (\cos B + \sin A)(\cos B - \sin A)\\
&= (\sin A + \cos B)[(\sin A + \cos B) + (\cos B - \sin A)]\\
&= (\sin A + \cos B)(2\cos B)\\
&= 2\cos B(\sin A + \cos B)\\
&\equiv \text{RHS.}
\end{aligned}$$

$$\frac{\cosec A - 1}{\cosec A + 1} \equiv \frac{1 - \sin A}{1 + \sin A}$$

Eyeballing and Mental Gymnastics

1. $\cosec = 1/\sin$
2. *rearrange and simplify.*

$$\text{LHS} = \frac{\cosec A - 1}{\cosec A + 1}$$

$$= \frac{\dfrac{1}{\sin A} - 1}{\dfrac{1}{\sin A} + 1}$$

$$= \frac{1 - \sin A}{\sin A} \cdot \frac{\sin A}{1 + \sin A}$$

$$= \frac{1 - \sin A}{1 + \sin A}$$

$$\equiv \text{RHS.}$$

$$\boxed{\frac{1+\tan A}{1-\tan A} \equiv \frac{\cot A + 1}{\cot A - 1}}$$

Eyeballing and Mental Gymnastics

1. $t = s/c$, $\cot = c/s$
2. *rearrange and simplify.*

$$\text{LHS} = \frac{1+\tan A}{1-\tan A}$$

$$= \frac{1 + \dfrac{\sin A}{\cos A}}{1 - \dfrac{\sin A}{\cos A}}$$

$$= \frac{\cos A + \sin A}{\cos A} \cdot \frac{\cos A}{\cos A - \sin A}$$

$$= \frac{\cos A + \sin A}{\cos A - \sin A}$$

$$= \frac{\cot A + 1}{\cot A - 1}$$

divide both numerator and denominator by sin

$$\equiv \text{RHS.}$$

$$3\sin^2 A + 4\cos^2 A \equiv 3 + \cos^2 A$$

Eyeballing and Mental Gymnastics

1. $s^2 + c^2 \equiv 1$
2. *rearrange and simplify.*

$$
\begin{aligned}
\text{LHS} &= 3\sin^2 A + 4\cos^2 A \\
&= 3\sin^2 A + 3\cos^2 A + \cos^2 A \\
&= 3(\sin^2 A + \cos^2 A) + \cos^2 A \\
&= 3 + \cos^2 A \\
&\equiv \text{RHS.}
\end{aligned}
$$

$$\boxed{\tan^2 A \cos^2 A + \cot^2 A \sin^2 A \equiv 1}$$

Eyeballing and Mental Gymnastics

1. $t = s/c$, $\cot = c/s$
2. $s^2 + c^2 \equiv 1$
3. *rearrange and simplify.*

$$
\begin{aligned}
\text{LHS} &= \tan^2 A \cos^2 A + \cot^2 A \sin^2 A \\
&= \frac{\sin^2 A}{\cos^2 A} \cdot \cos^2 A + \frac{\cos^2 A}{\sin^2 A} \cdot \sin^2 A \\
&= \sin^2 A + \cos^2 A \\
&= 1 \\
&\equiv \text{RHS.}
\end{aligned}
$$

$$\boxed{\frac{\sec A - \operatorname{cosec} A}{\sec A \operatorname{cosec} A} \equiv \sin A - \cos A}$$

Eyeballing and Mental Gymnastics

1. $\sec = 1/\cos$, $\operatorname{cosec} = 1/\sin$
2. *rearrange and simplify.*

$$
\begin{aligned}
\text{LHS} &= \frac{\sec A - \operatorname{cosec} A}{\sec A \operatorname{cosec} A} \\[2mm]
&= \frac{\sec A}{\sec A \operatorname{cosec} A} - \frac{\operatorname{cosec} A}{\sec A \operatorname{cosec} A} \\[2mm]
&= \frac{1}{\operatorname{cosec} A} - \frac{1}{\sec A} \\[2mm]
&= \sin A - \cos A \\[2mm]
&\equiv \text{RHS.}
\end{aligned}
$$

$$(\tan A + \tan B)(1 - \cot A \cot B) + (\cot A + \cot B)(1 - \tan A \tan B) \equiv 0$$

Eyeballing and Mental Gymnastics

1. $\cot = 1/\tan$*
2. *expansion of factors*
3. *rearrange and simplify.*

$$\text{LHS} = (\tan A + \tan B)(1 - \cot A \cot B) + (\cot A + \cot B)(1 - \tan A \tan B)$$

$$= \tan A + \tan B - \tan A \cot A \cot B - \tan B \cot A \cot B$$

$$+ \cot A + \cot B - \cot A \tan A \tan B - \cot B \tan A \tan B$$

$$= \tan A + \tan B - \cot B - \cot A$$

$$+ \cot A + \cot B - \tan B - \tan A$$

$$= 0$$

$$\equiv \text{RHS.}$$

---

*Since the terms in the identity to be proved are all tan and cot, it is easier to think of cot as 1/tan rather than to convert tan and cot into sin and cos.

$$\boxed{\frac{\cos A + \sin A - \sin^3 A}{\sin A} \equiv \cot A + \cos^2 A}$$

Eyeballing and Mental Gymnastics

1. $\sin A - \sin^3 A$ *suggests* $\sin A(1 - \sin^2 A)$
2. $s^2 + c^2 \equiv 1$
3. $\cot = c/s$
4. *rearrange and simplify.*

$$\begin{aligned}
\text{LHS} &= \frac{\cos A + \sin A - \sin^3 A}{\sin A} \\[2mm]
&= \frac{\cos A + \sin A(1 - \sin^2 A)}{\sin A} \\[2mm]
&= \frac{\cos A + \sin A(\cos^2 A)}{\sin A} \\[2mm]
&= \cot A + \cos^2 A \\[2mm]
&\equiv \text{RHS.}
\end{aligned}$$

$$\boxed{\frac{\cos^2 A - \sin^2 A}{1 - \tan^2 A} \equiv \cos^2 A}$$

Eyeballing and Mental Gymnastics

1. $t = s/c$
2. $s^2 + c^2 \equiv 1$
3. *rearrange and simplify.*

$$\text{LHS} = \frac{\cos^2 A - \sin^2 A}{1 - \tan^2 A}$$

$$= \frac{(\cos^2 A - \sin^2 A)}{1 - \dfrac{\sin^2 A}{\cos^2 A}}$$

$$= (\cos^2 A - \sin^2 A)\left(\frac{\cos^2 A}{\cos^2 A - \sin^2 A}\right)$$

$$= \cos^2 A$$

$$\equiv \text{RHS.}$$

$$\boxed{\cos^2 A(1 + \tan^2 A) \equiv 1}$$

Eyeballing and Mental Gymnastics

1. c^2, t^2 *suggest* $s^2 + c^2 = 1$
2. $t = s/c$
3. *rearrange and simplify.*

$$\text{LHS} = \cos^2 A(1 + \tan^2 A)$$

$$= \cos^2 A + \cos^2 A \cdot \frac{\sin^2 A}{\cos^2 A}$$

$$= \cos^2 A + \sin^2 A$$

$$= 1$$

$$\equiv \text{RHS.}$$

$$\cos^2 A - \sin^2 A \equiv 2\cos^2 A - 1$$

Eyeballing and Mental Gymnastics

1. c^2, s^2 suggest $s^2 + c^2 \equiv 1$
2. *simplify.*

$$\text{LHS} = \cos^2 A - \sin^2 A$$
$$= \cos^2 A - (1 - \cos^2 A)$$
$$= \cos^2 A - 1 + \cos^2 A$$
$$= 2\cos^2 A - 1$$
$$\equiv \text{RHS.}$$

$$\boxed{\cos^2 A + \tan^2 A \cos^2 A \equiv 1}$$

Eyeballing and Mental Gymnastics

1. c^2, t^2 *suggest* $s^2 + c^2 \equiv 1$
2. $t = s/c$
3. *rearrange and simplify.*

$$\text{LHS} = \cos^2 A + \tan^2 A \cos^2 A$$

$$= \cos^2 A + \frac{\sin^2 A}{\cos^2 A} \cdot \cos^2 A$$

$$= \cos^2 A + \sin^2 A$$

$$= 1$$

$$\equiv \text{RHS}.$$

$$\tan A \cos A + \operatorname{cosec} A \sin^2 A \equiv 2 \sin A$$

Eyeballing and Mental Gymnastics

1. $t = s/c$, $\operatorname{cosec} = 1/\sin$
2. *rearrange and simplify.*

$$\text{LHS} = \tan A \cos A + \operatorname{cosec} A \sin^2 A$$

$$= \frac{\sin A}{\cos A} \cdot \cos A + \frac{1}{\sin A} \cdot \sin^2 A$$

$$= \sin A + \sin A$$

$$= 2 \sin A$$

$$= \text{RHS.}$$

$$\boxed{\cos A(\sec A - \cos A) \equiv \sin^2 A}$$

Eyeballing and Mental Gymnastics

1. $\sec = 1/\cos$
2. *rearrange and simplify.*

$$\begin{aligned}
\text{LHS} &= \cos A(\sec A - \cos A) \\
&= \cos A\left(\frac{1}{\cos A} - \cos A\right) \\
&= 1 - \cos^2 A \\
&= \sin^2 A \\
&\equiv \text{RHS.}
\end{aligned}$$

$$\boxed{(\cos^2 A - 1)(\tan^2 A + 1) \equiv -\tan^2 A}$$

Eyeballing and Mental Gymnastics

1. c^2, t^2 suggest $s^2 + c^2 \equiv 1$
2. $t = s/c$
3. *rearrange and simplify.*

$$\text{LHS} = (\cos^2 A - 1)(\tan^2 A + 1)$$

$$= -\sin^2 A \cdot \sec^2 A$$

$$= -\frac{\sin^2 A}{\cos^2 A}$$

$$= -\tan^2 A$$

$$\equiv \text{RHS.}$$

$$\boxed{\frac{\cot A \cos A}{\operatorname{cosec}^2 - 1} \equiv \sin A}$$

Eyeballing and Mental Gymnastics

1. cosec^2 *suggests* $s^2 + c^2 \equiv 1$
2. $\cot = c/s$
3. *rearrange and simplify.*

$$\text{LHS} = \frac{\cot A \cos A}{(\operatorname{cosec}^2 - 1)}$$

$$= \frac{\cos A}{\sin A} \cdot \cos A \left(\frac{1}{\cot^2 A}\right)$$

$$= \frac{\cos A}{\sin A} \cdot \cos A \cdot \frac{\sin^2 A}{\cos^2 A}$$

$$= \sin A$$

$$\equiv \text{RHS}.$$

$$\boxed{\dfrac{\sin A}{\operatorname{cosec} A} + \dfrac{\cos A}{\sec A} \equiv 1}$$

Eyeballing and Mental Gymnastics

1. $\operatorname{cosec} = 1/\sin;\ \sec = 1/\cos$
2. *rearrange and simplify.*

$$\text{LHS} = \dfrac{\sin A}{\operatorname{cosec} A} + \dfrac{\cos A}{\sec A}$$

$$= \sin A \cdot \sin A + \cos A \cdot \cos A$$

$$= \sin^2 A + \cos^2 A$$

$$= 1$$

$$\equiv \text{RHS.}$$

$$\boxed{\frac{\cos^2 A}{1 - \sin A} \equiv 1 + \sin A}$$

Eyeballing and Mental Gymnastics

1. c^2 suggests $s^2 + c^2 \equiv 1$
2. *rearrange and simplify.*

$$\text{LHS} = \frac{\cos^2 A}{1 - \sin A}$$

$$= \frac{1 - \sin^2 A}{1 - \sin A}$$

$$= \frac{(1 + \sin A)(1 - \sin A)}{(1 - \sin A)}$$

$$= 1 + \sin A$$

$$\equiv \text{RHS.}$$

$$\boxed{\frac{\cos A \,\cosec A}{\tan A} \equiv \cot^2 A}$$

Eyeballing and Mental Gymnastics

1. $\cosec = 1/\sin$, $t = s/c$, $\cot = c/s$
2. *simplify.*

$$\text{LHS} = \frac{\cos A \,\cosec A}{\tan A}$$

$$= \cos A \cdot \frac{1}{\sin A} \cdot \frac{\cos A}{\sin A}$$

$$= \frac{\cos^2 A}{\sin^2 A}$$

$$= \cot^2 A$$

$$\equiv \text{RHS}.$$

$$\boxed{\frac{\tan A \sin A}{\sec^2 A - 1} \equiv \cos A}$$

Eyeballing and Mental Gymnastics

1. $\sec^2$ *suggest* $s^2 + c^2 \equiv 1$
2. *simplify.*

$$\text{LHS} = \frac{\tan A \sin A}{\sec^2 A - 1}$$

$$= \frac{\sin A}{\cos A} \cdot \sin A \cdot \frac{1}{\tan^2 A}$$

$$= \frac{\sin^2 A}{\cos A} \cdot \frac{\cos^2 A}{\sin^2 A}$$

$$= \cos A$$

$$\equiv \text{RHS.}$$

$$\boxed{1 + 2\sec^2 A \equiv 2\tan^2 A + 3}$$

Eyeballing and Mental Gymnastics

1. $\sec^2$, $\tan^2$ *suggests* $s^2 + c^2 \equiv 1$
2. *rearrange and simplify.*

$$\text{LHS} = 1 + 2\sec^2 A$$

$$= 1 + 2(1 + \tan^2 A)$$

$$= 1 + 2 + 2\tan^2 A$$

$$= 2\tan^2 A + 3$$

$$= \text{RHS.}$$

$$\frac{\cos A}{1 - \sin^2 A - \cos^2 A + \sin A} \equiv \cot A$$

Eyeballing and Mental Gymnastics

1. s^2, c^2 *suggest* $s^2 + c^2 \equiv 1$
2. $\cot = c/s$
3. *rearrange and simplify.*

$$\text{LHS} = \frac{\cos A}{1 - \sin^2 A - \cos^2 A + \sin A}$$

$$= \frac{\cos A}{1 - (\sin^2 A + \cos^2 A) + \sin A}$$

$$= \frac{\cos A}{\sin A}$$

$$= \cot A$$

$$\equiv \text{RHS.}$$

$$\boxed{\frac{1-\sin^2 A}{1-\cos^2 A} + \tan A \cot A \equiv \operatorname{cosec}^2 A}$$

Eyeballing and Mental Gymnastics

1. s^2, c^2 *suggest* $s^2 + c^2 \equiv 1$
2. $\tan \cdot \cot = 1$
3. *rearrange and simplify.*

$$\text{LHS} = \frac{1-\sin^2 A}{1-\cos^2 A} + \tan A \cot A$$

$$= \frac{\cos^2 A}{\sin^2 A} + 1$$

$$= \cot^2 A + 1$$

$$= \operatorname{cosec}^2 A$$

$$\equiv \text{RHS.}$$

$$\boxed{\operatorname{cosec} A + \cot A + \tan A \equiv \frac{1 + \cos A}{\sin A \cos A}}$$

Eyeballing and Mental Gymnastics

1. $\operatorname{cosec} = 1/\sin$, $\cot = c/s$, $t = s/c$
2. *rearrange and simplify.*

$$
\begin{aligned}
\text{LHS} &= \operatorname{cosec} A + \cot A + \tan A \\
&= \frac{1}{\sin A} + \frac{\cos A}{\sin A} + \frac{\sin A}{\cos A} \\
&= \frac{\cos A + \cos^2 A + \sin^2 A}{\sin A \cos A} \\
&= \frac{1 + \cos A}{\sin A \cos A} \\
&\equiv \text{RHS.}
\end{aligned}
$$

$$\frac{\cos^2 A}{\sin^2 A} + \cos^2 A + \sin^2 A \equiv \frac{1}{\sin^2 A}$$

Eyeballing and Mental Gymnastics

1. s^2, c^2 suggests $s^2 + c^2 \equiv 1$
2. *rearrange and simplify.*

$$\text{LHS} = \frac{\cos^2 A}{\sin^2 A} + \cos^2 A + \sin^2 A$$

$$= \frac{\cos^2 A}{\sin^2 A} + 1$$

$$= \frac{\cos^2 A + \sin^2 A}{\sin^2 A}$$

$$= \frac{1}{\sin^2 A}$$

$$\equiv \text{RHS.}$$

$$\boxed{(\cos A - \sin A)^2 + 2 \sin A \cos A \equiv 1}$$

Eyeballing and Mental Gymnastics

1. ( )^2 *suggests* $s^2 + c^2 \equiv 1$
2. *rearrange and simplify.*

$$\text{LHS} = (\cos A - \sin A)^2 + 2 \sin A \cos A$$

$$= (\cos^2 A - 2 \sin A \cos A + \sin^2 A) + 2 \sin A \cos A$$

$$= \cos^2 A + \sin^2 A$$

$$= 1$$

$$\equiv \text{RHS}.$$

$$\boxed{\sin A(\operatorname{cosec} A - \sin A) \equiv \cos^2 A}$$

Eyeballing and Mental Gymnastics

1. $\operatorname{cosec} = 1/\sin$
2. c^2 *suggests* $s^2 + c^2 = 1$
3. *simplify.*

$$\text{LHS} = \sin A(\operatorname{cosec} A - \sin A)$$

$$= \sin A \left(\frac{1}{\sin A} - \sin A \right)$$

$$= 1 - \sin^2 A$$

$$= \cos^2 A$$

$$\equiv \text{RHS.}$$

$$\boxed{\operatorname{cosec} A \sec A \equiv \tan A + \cot A}$$

Eyeballing and Mental Gymnastics

1. $\operatorname{cosec} = 1/\sin$, $\sec = 1/\cos$, $t = s/c$, $\cot = c/s$
2. *rearrange and simplify.*

$$\begin{aligned}
\text{LHS} &= \operatorname{cosec} A \sec A \\
&= \frac{1}{\sin A} \cdot \frac{1}{\cos A} \\
&= \frac{(\sin^2 A + \cos^2 A)}{\sin A \cos A} \\
&= \tan A + \cot A \\
&\equiv \text{RHS.}
\end{aligned}$$

$$\boxed{\frac{(2\sin^2 A - 1)^2}{\sin^4 A - \cos^4 A} \equiv 1 - 2\cos^2 A}$$

Eyeballing and Mental Gymnastics

1. $\sin^4 - \cos^4$ *suggests* $(a^2 - b^2)(a^2 + b^2)$
2. $2\sin^2 A - 1$ *equals* $1 - 2\cos^2 A$
3. *rearrange and simplify.*

$$\text{LHS} = \frac{(2\sin^2 A - 1)^2}{\sin^4 A - \cos^4 A}$$

$$= \frac{(1 - 2\cos^2 A)^2}{(\sin^2 A - \cos^2 A)(\sin^2 A + \cos^2 A)}$$

$$= \frac{(1 - 2\cos^2 A)^2}{(1 - 2\cos^2 A)(1)}$$

$$= 1 - 2\cos^2 A$$

$$\equiv \text{RHS.}$$

$$\boxed{\cos A + \tan A \sin A \equiv \sec A}$$

Eyeballing and Mental Gymnastics

1. $t = s/c$, $\sec = 1/\cos$
2. *simplify.*

$$\begin{aligned}
\text{LHS} &= \cos A + \tan A \sin A \\
&= \cos A + \frac{\sin A}{\cos A} \cdot \sin A \\
&= \frac{\cos^2 A + \sin^2 A}{\cos A} \\
&= \frac{1}{\cos A} \\
&= \sec A \\
&\equiv \text{RHS.}
\end{aligned}$$

$$\boxed{\sin A + \cos A \cot A \equiv \operatorname{cosec} A}$$

Eyeballing and Mental Gymnastics

1. $\cot = c/s$, $\operatorname{cosec} = 1/\sin$
2. *rearrange and simplify.*

$$
\begin{aligned}
\text{LHS} &= \sin A + \cos A \cot A \\
&= \sin A + \cos A \cdot \frac{\cos A}{\sin A} \\
&= \frac{\sin^2 A + \cos^2 A}{\sin A} \\
&= \frac{1}{\sin A} \\
&= \operatorname{cosec} A \\
&\equiv \text{RHS.}
\end{aligned}
$$

$$\boxed{\sec A - \sin A \tan A \equiv \cos A}$$

Eyeballing and Mental Gymnastics

1. $\sec = 1/\cos$, $t = s/c$
2. *rearrange and simplify.*

$$
\begin{aligned}
\text{LHS} &= \sec A - \sin A \tan A \\
&= \frac{1}{\cos A} - \frac{\sin A \cdot \sin A}{\cos A} \\
&= \frac{1 - \sin^2 A}{\cos A} \\
&= \frac{\cos^2 A}{\cos A} \\
&= \cos A \\
&\equiv \text{RHS.}
\end{aligned}
$$

$$\boxed{\operatorname{cosec} A - \sin A \equiv \cot A \cos A}$$

Eyeballing and Mental Gymnastics

1. $\operatorname{cosec} = 1/\sin$, $\cot = c/s$
2. *rearrange and simplify.*

$$\text{LHS} = \operatorname{cosec} A - \sin A$$

$$= \frac{1}{\sin A} - \sin A$$

$$= \frac{1 - \sin^2 A}{\sin A}$$

$$= \frac{\cos^2 A}{\sin A}$$

$$= \cot A \cdot \cos A$$

$$\equiv \text{RHS.}$$

$$(\cos A + \cot A)\sec A \equiv 1 + \operatorname{cosec} A$$

Eyeballing and Mental Gymnastics

1. $\cot = c/s$, $\sec = 1/\cos$, $\operatorname{cosec} = 1/\sin$
2. *rearrange and simplify.*

$$\mathrm{LHS} = (\cos A + \cot A)\sec A$$

$$= \left(\cos A + \frac{\cos A}{\sin A}\right) \cdot \frac{1}{\cos A}$$

$$= \cos A \left(1 + \frac{1}{\sin A}\right)\frac{1}{\cos A}$$

$$= 1 + \frac{1}{\sin A}$$

$$= 1 + \operatorname{cosec} A$$

$$\equiv \mathrm{RHS}.$$

$$\boxed{\operatorname{cosec} A - \sin A \equiv \cos A \cot A}$$

Eyeballing and Mental Gymnastics

1. $\operatorname{cosec} = 1/\sin$, $\cot = c/s$
2. *rearrange and simplify.*

$$\text{LHS} = \operatorname{cosec} A - \sin A$$

$$= \frac{1}{\sin A} - \sin A$$

$$= \frac{1 - \sin^2 A}{\sin A}$$

$$= \frac{\cos^2 A}{\sin A}$$

$$= \cos A \cot A$$

$$\equiv \text{RHS}.$$

$$\boxed{\sec A - \cos A \equiv \sin A \tan A}$$

Eyeballing and Mental Gymnastics

1. $\sec = 1/\cos$, $t = s/c$
2. *rearrange and simplify.*

$$
\begin{aligned}
\text{LHS} &= \sec A - \cos A \\
&= \frac{1}{\cos A} - \cos A \\
&= \frac{1 - \cos^2 A}{\cos A} \\
&= \frac{\sin^2 A}{\cos A} \\
&= \sin A \tan A \\
&\equiv \text{RHS.}
\end{aligned}
$$

$$\boxed{\tan A + \cot A \equiv \sec A \, \text{cosec} \, A}$$

Eyeballing and Mental Gymnastics

1. $t = s/c$, $\cot = c/s$, $\sec = 1/\cos$, $\text{cosec} = 1/\sin$
2. $s^2 + c^2 \equiv 1$
3. *rearrange and simplify.*

$$
\begin{aligned}
\text{LHS} &= \tan A + \cot A \\[1mm]
&= \frac{\sin A}{\cos A} + \frac{\cos A}{\sin A} \\[2mm]
&= \frac{\sin^2 A + \cos^2 A}{\sin A \cos A} \\[2mm]
&= \frac{1}{\sin A \cos A} \\[2mm]
&= \sec A \, \text{cosec} \, A \\[2mm]
&\equiv \text{RHS.}
\end{aligned}
$$

$$\boxed{\operatorname{cosec}^2 A - \cot^2 A \equiv 1}$$

Eyeballing and Mental Gymnastics

1. $\operatorname{cosec} = 1/\sin$, $\cot = c/s$
2. cosec^2, $\cot^2$ *suggest* $s^2 + c^2 \equiv 1$
3. *simplify.*

$$
\begin{aligned}
\text{LHS} &= \operatorname{cosec}^2 A - \cot^2 A \\[2mm]
&= \frac{1}{\sin^2 A} - \frac{\cos^2 A}{\sin^2 A} \\[2mm]
&= \frac{1 - \cos^2 A}{\sin^2 A} \\[2mm]
&= \frac{\sin^2 A}{\sin^2 A} \\[2mm]
&= 1 \\[2mm]
&\equiv \text{RHS.}
\end{aligned}
$$

$$\boxed{\tan^2 A + 1 \equiv \sec^2 A}$$

Eyeballing and Mental Gymnastics

1. $\tan^2$, $\sec^2$ *suggest* $s^2 + c^2 \equiv 1$
2. *simplify.*

$$\text{LHS} = \tan^2 A + 1$$

$$= \frac{\sin^2 A}{\cos^2 A} + 1$$

$$= \frac{\sin^2 A + \cos^2 A}{\cos^2 A}$$

$$= \frac{1}{\cos^2 A}$$

$$= \sec^2 A$$

$$\equiv \text{RHS.}$$

$$\boxed{\cos A + \cos A \cot^2 A = \cot A \operatorname{cosec} A}$$

Eyeballing and Mental Gymnastics

1. c^2 *suggests* $s^2 + c^2 \equiv 1$
2. $\operatorname{cosec} = 1/\sin$
3. *rearrange and simplify.*

$$\mathrm{LHS} = \cos A + \cos A \cot^2 A$$

$$= \cos A (1 + \cot^2 A)$$

$$= \cos A \cdot \operatorname{cosec}^2 A$$

$$= \cos A \cdot \frac{1}{\sin^2 A}$$

$$= \cot A \operatorname{cosec} A$$

$$= \mathrm{RHS}.$$

$$\boxed{\cot A(\sec^2 A - 1) \equiv \tan A}$$

Eyeballing and Mental Gymnastics

1. $\cot = c/s$, $\sec = 1/\cos$, $t = s/c$
2. $\sec^2$ *suggests* $s^2 + c^2 \equiv 1$
3. *rearrange and simplify.*

$$\text{LHS} = \cot A(\sec^2 A - 1)$$

$$= \frac{1}{\tan A}(\tan^2 A)$$

$$= \tan A$$

$$\equiv \text{RHS.}$$

$$\boxed{1 - 2\cos^2 A \equiv 2\sin^2 A - 1}$$

Eyeballing and Mental Gymnastics

1. c^2, s^2, 1 suggest $s^2 + c^2 \equiv 1$
2. *rearrange and simplify.*

$$\begin{aligned}
\text{LHS} &= 1 - 2\cos^2 A \\
&= 1 - 2(1 - \sin^2 A) \\
&= 1 - 2 + 2\sin^2 A \\
&= 2\sin^2 A - 1 \\
&\equiv \text{RHS.}
\end{aligned}$$

$$\cos^2 A(\operatorname{cosec}^2 A - \cot^2 A) \equiv \cos^2 A$$

Eyeballing and Mental Gymnastics

1. $\operatorname{cosec} = 1/\sin$, $\cot = c/s$
2. $\cos^2$, cosec^2, $\cot^2$ *suggest* $s^2 + c^2 \equiv 1$
3. *rearrange and simplify.*

$$\text{LHS} = \cos^2 A(\operatorname{cosec}^2 A - \cot^2 A)$$

$$= \cos^2 A \left(\frac{1}{\sin^2 A} - \frac{\cos^2 A}{\sin^2 A} \right)$$

$$= \cos^2 A \left(\frac{1 - \cos^2 A}{\sin^2 A} \right)$$

$$= \cos^2 A \left(\frac{\sin^2 A}{\sin^2 A} \right)$$

$$= \cos^2 A$$

$$\equiv \text{RHS}.$$

$$\boxed{\tan^2 A - \sin^2 A \equiv \tan^2 A \sin^2 A}$$

Eyeballing and Mental Gymnastics

1. $t = s/c$
2. t^2, s^2 suggest $s^2 + c^2 \equiv 1$
3. *rearrange and simplify.*

$$\text{LHS} = \tan^2 A - \sin^2 A$$

$$= \frac{\sin^2 A}{\cos^2 A} - \sin^2 A$$

$$= \frac{\sin^2 A - \sin^2 A \cos^2 A}{\cos^2 A}$$

$$= \frac{\sin^2 A (1 - \cos^2 A)}{\cos^2 A}$$

$$= \tan^2 A \cdot \sin^2 A$$

$$\equiv \text{RHS}.$$

$$\boxed{\frac{1+\cot A}{1+\tan A} \equiv \cot A}$$

Eyeballing and Mental Gymnastics

1. $\cot = c/s, t = s/c$
2. *rearrange and simplify.*

$$\text{LHS} = \frac{1+\cot A}{1+\tan A}$$

$$= \left(1 + \frac{\cos A}{\sin A}\right)\left(\frac{1}{1 + \dfrac{\sin A}{\cos A}}\right)$$

$$= \left(\frac{\sin A + \cos A}{\sin A}\right) \cdot \left(\frac{\cos A}{\cos A + \sin A}\right)$$

$$= \frac{\cos A}{\sin A}$$

$$= \cot A$$

$$\equiv \text{RHS.}$$

$$(\cos^2 A - 2)^2 - 4\sin^2 A \equiv \cos^4 A$$

Eyeballing and Mental Gymnastics

1. c^2, s^2 suggest $s^2 + c^2 \equiv 1$
2. *rearrange and simplify.*

$$\begin{aligned}
\text{LHS} &= (\cos^2 A - 2)^2 - 4\sin^2 A \\
&= (\cos^4 A - 4\cos^2 A + 4) - 4\sin^2 A \\
&= \cos^4 A - 4(\cos^2 A + \sin^2 A) + 4 \\
&= \cos^4 A - 4(1) + 4 \\
&= \cos^4 A \\
&\equiv \text{RHS.}
\end{aligned}$$

$$\boxed{\sin A + \sin A \tan^2 A \equiv \tan A \sec A}$$

Eyeballing and Mental Gymnastics

1. t^2 suggests $s^2 + c^2 \equiv 1$
2. $\sec = 1/\cos$
3. *rearrange and simplify.*

$$\begin{aligned}
\text{LHS} &= \sin A + \sin A \tan^2 A \\
&= \sin A (1 + \tan^2 A) \\
&= \sin A \cdot \sec^2 A \\
&= \sin A \cdot \frac{1}{\cos^2 A} \\
&= \tan A \sec A \\
&\equiv \text{RHS.}
\end{aligned}$$

$$\boxed{1 - 2\sin^2 A \equiv 2\cos^2 A - 1}$$

Eyeballing and Mental Gymnastics

1. s^2, c^2 suggest $s^2 + c^2 \equiv 1$
2. *rearrange and simplify.*

$$\begin{aligned}
\text{LHS} &= 1 - 2\sin^2 A \\
&= 1 - \sin^2 A - \sin^2 A \\
&= \cos^2 A - \sin^2 A \\
&= \cos^2 A + (\cos^2 A - 1) \\
&= 2\cos^2 A - 1 \\
&\equiv \text{RHS.}
\end{aligned}$$

$$\boxed{(\sin A + \cos A)^2 + (\sin A - \cos A)^2 \equiv 2}$$

Eyeballing and Mental Gymnastics

1. $(\ \ )^2$ *suggests expansion*
2. $s^2 + c^2 \equiv 1$
3. *rearrange and simplify.*

$$\begin{aligned}
\text{LHS} &= (\sin A + \cos A)^2 + (\sin A - \cos A)^2 \\
&= (\sin^2 A + 2\sin A \cos A + \cos^2 A) \\
&\quad + (\sin^2 A - 2\sin A \cos A + \cos^2 A) \\
&= 2(\sin^2 A + \cos^2 A) \\
&= 2(1) \\
&= 2 \\
&\equiv \text{RHS.}
\end{aligned}$$

$$\boxed{\sec^4 A - \sec^2 A \equiv \tan^4 A + \tan^2 A}$$

Eyeballing and Mental Gymnastics

1. $\sec = 1/\cos$, $t = s/c$
2. $s^2 + c^2 \equiv 1$
3. *rearange and simplify.*

$$\begin{aligned}
\text{LHS} &= \sec^4 A - \sec^2 A \\
&= \sec^2 A (\sec^2 A - 1) \\
&= \sec^2 A \tan^2 A \\
&= (\tan^2 A + 1) \tan^2 A \\
&= \tan^4 A + \tan^2 A \\
&\equiv \text{RHS.}
\end{aligned}$$

$$\boxed{\operatorname{cosec}^4 A - \operatorname{cosec}^2 A \equiv \cot^4 A + \cot^2 A}$$

Eyeballing and Mental Gymnastics

1. $\operatorname{cosec} = 1/\sin, \cot = c/s$
2. $s^2 + c^2 \equiv 1$
3. *rearrange and simplify.*

$$\text{LHS} = \operatorname{cosec}^4 A - \operatorname{cosec}^2 A$$

$$= \operatorname{cosec}^2 A(\operatorname{cosec}^2 A - 1)$$

$$= \operatorname{cosec}^2 A(\cot^2 A)$$

$$= (1 + \cot^2 A)(\cot^2 A)$$

$$= \cot^2 A + \cot^4 A$$

$$\equiv \text{RHS}.$$

$$\boxed{\frac{\operatorname{cosec} A}{\cot A + \tan A} \equiv \cos A}$$

Eyeballing and Mental Gymnastics

1. $\operatorname{cosec} = 1/\sin,\ \cot = c/s,\ t = s/c$
2. *rearrange and simplify.*

$$\text{LHS} = \frac{\operatorname{cosec} A}{\cot A + \tan A}$$

$$= \frac{1}{\sin A}\left(\frac{1}{\dfrac{\cos A}{\sin A} + \dfrac{\sin A}{\cos A}}\right)$$

$$= \frac{1}{\sin A}\left(\frac{\sin A \cos A}{\cos^2 A + \sin^2 A}\right)$$

$$= \frac{1}{\sin A}\left(\frac{\sin A \cos A}{1}\right)$$

$$= \cos A$$

$$\equiv \text{RHS}.$$

$$\boxed{\frac{1 + \sec A}{1 + \cos A} \equiv \sec A}$$

Eyeballing and Mental Gymnastics

1. $\sec = 1/\cos$
2. *rearrange and simplify.*

$$\text{LHS} = \frac{1 + \sec A}{1 + \cos A}$$

$$= \left(1 + \frac{1}{\cos A}\right)\left(\frac{1}{1 + \cos A}\right)$$

$$= \left(\frac{\cos A + 1}{\cos A}\right)\left(\frac{1}{1 + \cos A}\right)$$

$$= \frac{1}{\cos A}$$

$$= \sec A$$

$$\equiv \text{RHS.}$$

$$\boxed{\dfrac{\sin A}{1 + \cos A} + \dfrac{1 + \cos A}{\sin A} \equiv \dfrac{2}{\sin A}}$$

Eyeballing and Mental Gymnastics

1. *common denominator*
2. $s^2 + c^2 \equiv 1$
3. *rearrange and simplify.*

$$\text{LHS} = \frac{\sin A}{1 + \cos A} + \frac{1 + \cos A}{\sin A}$$

$$= \frac{\sin^2 A + (1 + \cos A)^2}{(1 + \cos A)(\sin A)}$$

$$= \frac{\sin^2 A + (1 + 2\cos A + \cos^2 A)}{(1 + \cos A)(\sin A)}$$

$$= \frac{2 + 2\cos A}{(1 + \cos A)(\sin A)}$$

$$= \frac{2}{\sin A}$$

$$\equiv \text{RHS.}$$

$$\boxed{\frac{\sec A + \csc A}{1 + \tan A} \equiv \csc A}$$

Eyeballing and Mental Gymnastics

1. $\sec = 1/\cos$, $\csc = 1/\sin$, $t = s/c$
2. *rearrange and simplify.*

$$\text{LHS} = \frac{\sec A + \csc A}{1 + \tan A}$$

$$= \left(\frac{1}{\cos A} + \frac{1}{\sin A} \right) \left(\frac{1}{1 + \dfrac{\sin A}{\cos A}} \right)$$

$$= \frac{(\sin A + \cos A)}{\cos A \sin A} \left(\frac{\cos A}{\cos A + \sin A} \right)$$

$$= \frac{1}{\sin A}$$

$$= \csc A$$

$$\equiv \text{RHS.}$$

$$\boxed{\frac{1}{\tan A + \cot A} \equiv \frac{\sin A}{\sec A}}$$

Eyeballing and Mental Gymnastics

1. $t = s/c$, $\cot = c/s$, $\sec = 1/\cos$
2. *common denominator*
3. *rearrange and simplify.*

$$\text{LHS} = \frac{1}{\tan A + \cot A}$$

$$= \frac{1}{\dfrac{\sin A}{\cos A} + \dfrac{\cos A}{\sin A}}$$

$$= \frac{\cos A \sin A}{\sin^2 A + \cos^2 A}$$

$$= \cos A \sin A$$

$$= \frac{\sin A}{\sec A}$$

$$\equiv \text{RHS.}$$

$$\frac{1+\sin A}{\cos A} + \frac{\cos A}{1+\sin A} \equiv \frac{2}{\cos A}$$

Eyeballing and Mental Gymnastics

1. *common denominator*
2. $s^2 + c^2 \equiv 1$
3. *rearrange and simplify.*

$$\text{LHS} = \frac{1+\sin A}{\cos A} + \frac{\cos A}{1+\sin A}$$

$$= \frac{(1+\sin A)^2 + \cos^2 A}{\cos A(1+\sin A)}$$

$$= \frac{1+2\sin A + \sin^2 A + \cos^2 A}{\cos A(1+\sin A)}$$

$$= \frac{2+2\sin A}{\cos A(1+\sin A)}$$

$$= \frac{2}{\cos A}$$

$$\equiv \text{RHS.}$$

$$\boxed{\frac{1 - \cos A}{\sin A} \equiv \frac{\sin A}{1 + \cos A}}$$

Eyeballing and Mental Gymnastics

1. 1 *suggests* $s^2 + c^2 \equiv 1$
2. *rearrange and simplify.*

$$\begin{aligned}
\text{LHS} &= \frac{1 - \cos A}{\sin A} \\[2mm]
&= \frac{1 - \cos A}{\sin A} \cdot \left(\frac{1 + \cos A}{1 + \cos A} \right) \\[2mm]
&= \frac{1 - \cos^2 A}{\sin A (1 + \cos A)} \\[2mm]
&= \frac{\sin^2 A}{\sin A (1 + \cos A)} \\[2mm]
&= \frac{\sin A}{1 + \cos A} \\[2mm]
&\equiv \text{RHS.}
\end{aligned}$$

$$\frac{1}{\tan A + \cot A} \equiv \sin A \cos A$$

Eyeballing and Mental Gymnastics

1. $t = s/c$, $\cot = c/s$
2. *rearrange and simplify.*

$$\text{LHS} = \frac{1}{\tan A + \cot A}$$

$$= \frac{1}{\dfrac{\sin A}{\cos A} + \dfrac{\cos A}{\sin A}}$$

$$= \frac{\cos A \sin A}{\sin^2 A + \cos^2 A}$$

$$= \frac{\sin A \cos A}{1}$$

$$= \sin A \cos A$$

$$\equiv \text{RHS.}$$

$$(\operatorname{cosec} A - \sin A)(\sec A - \cos A)(\tan A + \cot A) \equiv 1$$

Eyeballing and Mental Gymnastics

1. $\operatorname{cosec} = 1/\sin$, $\sec = 1/\cos$, $t = s/c$, $\cot = c/s$
2. *common denominator*
3. *rearrange and simplify.*

$$\text{LHS} = (\operatorname{cosec} A - \sin A)(\sec A - \cos A)(\tan A + \cot A)$$

$$= \left(\frac{1}{\sin A} - \sin A \right) \left(\frac{1}{\cos A} - \cos A \right) \left(\frac{\sin A}{\cos A} + \frac{\cos A}{\sin A} \right)$$

$$= \left(\frac{(1 - \sin^2 A)}{\sin A} \right) \left(\frac{1 - \cos^2 A}{\cos A} \right) \left(\frac{\sin^2 A + \cos^2 A}{\cos A \sin A} \right)$$

$$= \left(\frac{\cos^2 A}{\sin A} \right) \left(\frac{\sin^2 A}{\cos A} \right) \left(\frac{1}{\cos A \sin A} \right)$$

$$= 1$$

$$\equiv \text{RHS.}$$

$$1 - \frac{\cos^2 A}{1 + \sin A} \equiv \sin A$$

Eyeballing and Mental Gymnastics

1. *Common denominator*
2. c^2 *suggests* $s^2 + c^2 = 1$
3. *rearrange and simplify.*

$$\text{LHS} = 1 - \frac{\cos^2 A}{1 + \sin A}$$

$$= \frac{1 + \sin A - \cos^2 A}{1 + \sin A}$$

$$= \frac{\sin^2 A + \sin A}{1 + \sin A}$$

$$= \frac{\sin A (1 + \sin A)}{(1 + \sin A)}$$

$$= \sin A$$

$$\equiv \text{RHS}.$$

$$\boxed{\frac{1 - 2\sin^2 A}{\sin A \cos A} \equiv \cot A - \tan A}$$

Eyeballing and Mental Gymnastics

1. $\cot = c/s, t = s/c$
2. s^2 *suggests* $s^2 + c^2 \equiv 1$
3. *rearrange and simplify.*

$$\text{LHS} = \frac{1 - 2\sin^2 A}{\sin A \cos A}$$

$$= \frac{1 - \sin^2 A - \sin^2 A}{\sin A \cos A}$$

$$= \frac{\cos^2 A - \sin^2 A}{\sin A \cos A}$$

$$= \frac{\cos A}{\sin A} - \frac{\sin A}{\cos A}$$

$$= \cot A - \tan A$$

$$\equiv \text{RHS.}$$

$$\boxed{\frac{\sin A + \cos A}{\sin A} - \frac{\cos A - \sin A}{\cos A} \equiv \operatorname{cosec} A \sec A}$$

Eyeballing and Mental Gymnastics

1. $\operatorname{cosec} = 1/\sin$, $\sec A = 1/\cos A$
2. *common denominator*
3. $s^2 + c^2 \equiv 1$
4. *rearrange and simplify.*

$$
\begin{aligned}
\text{LHS} &= \frac{\sin A + \cos A}{\sin A} - \frac{\cos A - \sin A}{\cos A} \\[2mm]
&= \frac{(\sin A \cos A + \cos^2 A) - (\sin A \cos A - \sin^2 A)}{\sin A \cos A} \\[2mm]
&= \frac{\cos^2 A + \sin^2 A}{\sin A \cos A} \\[2mm]
&= \frac{1}{\sin A \cos A} \\[2mm]
&= \operatorname{cosec} A \sec A \\[2mm]
&\equiv \text{RHS.}
\end{aligned}
$$

$$\boxed{\frac{\tan A + \tan B}{\cot A + \cot B} \equiv \tan A \tan B}$$

Eyeballing and Mental Gymnastics

1. $\cot = 1/\tan$
2. *since both LHS and RHS are in terms of* tan *and* cot, *it may be easier to leave the terms in* tan *rather than convert them to* sin *and* cos.
3. *common denominator*
4. *rearrange and simplify.*

$$\text{LHS} = \frac{\tan A + \tan B}{\cot A + \cot B}$$

$$= \frac{\tan A + \tan B}{\dfrac{1}{\tan A} + \dfrac{1}{\tan B}}$$

$$= \frac{(\tan A + \tan B)}{\left(\dfrac{\tan B + \tan A}{\tan A \tan B}\right)}$$

$$= (\tan A + \tan B)\left(\frac{\tan A \tan B}{\tan A + \tan B}\right)$$

$$= \tan A \tan B$$

$$\equiv \text{RHS}.$$

$$\frac{1+\sin A}{1-\sin A} - \frac{1-\sin A}{1+\sin A} \equiv 4\tan A \sec A$$

Eyeballing and Mental Gymnastics

1. $t = s/c$, $\sec = 1/\cos$
2. $s^2 + c^2 \equiv 1$
3. *rearrange and simplify.*

$$\text{LHS} = \frac{1+\sin A}{1-\sin A} - \frac{1-\sin A}{1+\sin A}$$

$$= \frac{(1+\sin A)^2 - (1-\sin A)^2}{(1-\sin A)(1+\sin A)}$$

$$= \frac{(1+2\sin A + \sin^2 A) - (1 - 2\sin A + \sin^2 A)}{1 - \sin^2 A}$$

$$= \frac{4\sin A}{\cos^2 A}$$

$$= 4\tan A \sec A$$

$$\equiv \text{RHS}.$$

$$\frac{\sin A + \cos A}{\cos A} - \frac{\sin A - \cos A}{\sin A} \equiv \sec A \, \text{cosec} \, A$$

Eyeballing and Mental Gymnastics

1. $\sec = 1/\cos$, $\text{cosec} = 1/\sin$
2. *common denominator*
3. $s^2 + c^2 \equiv 1$
4. *rearrange and simplify.*

$$\text{LHS} = \frac{\sin A + \cos A}{\cos A} - \frac{\sin A - \cos A}{\sin A}$$

$$= \frac{(\sin^2 A + \sin A \cos A) - (\sin A \cos A - \cos^2 A)}{\cos A \sin A}$$

$$= \frac{\sin^2 A + \cos^2 A}{\cos A \sin A}$$

$$= \frac{1}{\cos A \sin A}$$

$$= \sec A \, \text{cosec} \, A$$

$$\equiv \text{RHS.}$$

$$1 - \frac{\sin^2 A}{1 - \cos A} \equiv -\cos A$$

Eyeballing and Mental Gymnastics

1. $s^2 + c^2 \equiv 1$
2. *rearrange and simplify.*

$$\text{LHS} = 1 - \frac{\sin^2 A}{1 - \cos A}$$

$$= 1 - \frac{(1 - \cos^2 A)}{(1 - \cos A)}$$

$$= 1 - \frac{(1 - \cos A)(1 + \cos A)}{(1 - \cos A)}$$

$$= 1 - (1 + \cos A)$$

$$= -\cos A$$

$$\equiv \text{RHS}.$$

$$\boxed{\frac{1 - 2\cos^2 A}{\sin A \cos A} \equiv \tan A - \cot A}$$

Eyeballing and Mental Gymnastics

1. c^2 *suggests* $s^2 + c^2 \equiv 1$
2. $t = s/c$, $\cot = c/s$
3. *rearrange and simplify.*

$$\text{LHS} = \frac{1 - 2\cos^2 A}{\sin A \cos A}$$

$$= \frac{(\sin^2 A + \cos^2 A) - 2\cos^2 A}{\sin A \cos A}$$

$$= \frac{\sin^2 A - \cos^2 A}{\sin A \cos A}$$

$$= \frac{\sin A}{\cos A} - \frac{\cos A}{\sin A}$$

$$= \tan A - \cot A$$

$$\equiv \text{RHS}.$$

$$\frac{1}{1 - \cos A} + \frac{1}{1 + \cos A} \equiv 2 \operatorname{cosec}^2 A$$

Eyeballing and Mental Gymnastics

1. *common denominator*
2. $\operatorname{cosec} = 1/\sin$
3. cosec^2 *suggests* $s^2 + c^2 \equiv 1$
4. *rearrange and simplify.*

$$\text{LHS} = \frac{1}{1 - \cos A} + \frac{1}{1 + \cos A}$$

$$= \frac{1 + \cos A + 1 - \cos A}{(1 - \cos A)(1 + \cos A)}$$

$$= \frac{2}{1 - \cos^2 A}$$

$$= \frac{2}{\sin^2 A}$$

$$= 2 \operatorname{cosec}^2 A$$

$$\equiv \text{RHS.}$$

$$\boxed{\frac{1}{\sin A + 1} - \frac{1}{\sin A - 1} \equiv 2 \sec^2 A}$$

Eyeballing and Mental Gymnastics

1. $\sec = 1/\cos$
2. $\sec^2$ *suggests* $s^2 + c^2 \equiv 1$
3. *rearrange and simplify.*

$$
\begin{aligned}
\text{LHS} &= \frac{1}{\sin A + 1} - \frac{1}{\sin A - 1} \\[2mm]
&= \frac{(\sin A - 1) - (\sin A + 1)}{(\sin A + 1)(\sin A - 1)} \\[2mm]
&= \frac{-2}{\sin^2 A - 1} \\[2mm]
&= \frac{-2}{-\cos^2 A} \\[2mm]
&= 2 \sec^2 A \\[2mm]
&\equiv \text{RHS.}
\end{aligned}
$$

$$\boxed{\dfrac{\tan A(1+\cot^2 A)}{1+\tan^2 A} \equiv \cot A}$$

Eyeballing and Mental Gymnastics

1. $\cot^2$, $\tan^2$ *suggest* $s^2 + c^2 \equiv 1$
2. $\cot = c/s$, $t = s/c$
3. *rearrange and simplify.*

$$
\begin{aligned}
\text{LHS} &= \frac{\tan A(1+\cot^2 A)}{1+\tan^2 A} \\[2mm]
&= \frac{\sin A}{\cos A}\frac{(\mathrm{cosec}^2 A)}{(\sec^2 A)} \\[2mm]
&= \frac{\sin A}{\cos A}\left(\frac{1}{\sin^2 A}\right)\cdot\left(\frac{\cos^2 A}{1}\right) \\[2mm]
&= \frac{\cos A}{\sin A} \\[2mm]
&= \cot A \\[2mm]
&\equiv \text{RHS.}
\end{aligned}
$$

$$\boxed{\frac{1 - \sec^2 A}{(1 - \cos A)(1 + \cos A)} \equiv -\sec^2 A}$$

Eyeballing and Mental Gymnastics

1. *$\sec^2$ suggests $s^2 + c^2 \equiv 1$*
2. $\sec = 1/\cos$
3. *rearrange and simplify.*

$$\text{LHS} = \frac{1 - \sec^2 A}{(1 - \cos A)(1 + \cos A)}$$

$$= \frac{-\tan^2 A}{(1 - \cos^2 A)}$$

$$= -\frac{\sin^2 A}{\cos^2 A} \cdot \left(\frac{1}{\sin^2 A}\right)$$

$$= -\frac{1}{\cos^2 A}$$

$$= -\sec^2 A$$

$$\equiv \text{RHS.}$$

$$\boxed{\frac{\cot A(1+\tan^2 A)}{1+\cot^2 A} \equiv \tan A}$$

Eyeballing and Mental Gymnastics

1. $\tan^2$, $\cot^2$ *suggest* $s^2 + c^2 \equiv 1$
2. $\cot = c/s, t = s/c$
3. *rearrange and simplify.*

$$\text{LHS} = \frac{\cot A(1+\tan^2 A)}{(1+\cot^2 A)}$$

$$= \frac{\cos A}{\sin A}(\sec^2 A)\left(\frac{1}{\operatorname{cosec}^2 A}\right)$$

$$= \frac{\cos A}{\sin A}\cdot\left(\frac{1}{\cos^2 A}\right)\cdot\left(\frac{\sin^2 A}{1}\right)$$

$$= \frac{\sin A}{\cos A}$$

$$= \tan A$$

$$\equiv \text{RHS.}$$

$$\boxed{\frac{1}{\cos^2 A(1 + \tan^2 A)} \equiv 1}$$

Eyeballing and Mental Gymnastics

1. c^2, t^2 suggest $s^2 + c^2 = 1$
2. $t = s/c$
3. common denominator
4. rearrange and simplify.

$$\text{LHS} = \frac{1}{\cos^2 A(1 + \tan^2 A)}$$

$$= \frac{1}{\cos^2 A(\sec^2 A)}$$

$$= \frac{1}{\cos^2 A\left(\dfrac{1}{\cos^2 A}\right)}$$

$$= 1$$

$$\equiv \text{RHS}.$$

$$(\tan A + \cot A)^2 - (\tan A - \cot A)^2 \equiv 4$$

Eyeballing and Mental Gymnastics

1. $\cot = 1/\tan$
2. *expansion of squares*
3. $(\quad)^2$ *suggests* $s^2 + c^2 \equiv 1$
4. *rearrange and simplify.*

$\text{LHS} = (\tan A + \cot A)^2 - (\tan A - \cot A)^2$

$\quad = (\tan^2 A + 2\tan A \cot A + \cot^2 A) - (\tan^2 A - 2\tan A \cot A + \cot^2 A)$

$\quad = 2\tan A \cot A + 2\tan A \cot A$

$\quad = 2 + 2$

$\quad = 4$

$\quad \equiv \text{RHS.}$

$$\boxed{\frac{1 - \cot^2 A}{1 + \cot^2 A} = \sin^2 A - \cos^2 A}$$

Eyeballing and Mental Gymnastics

1. $\cot^2$, s^2, c^2 suggest $s^2 + c^2 \equiv 1$
2. $\cot = c/s$
3. *rearrange and simplify.*

$$\text{LHS} = \frac{1 - \cot^2 A}{1 + \cot^2 A}$$

$$= \frac{1 - \cot^2 A}{\text{cosec}^2 A}$$

$$= \left(1 - \frac{\cos^2 A}{\sin^2 A}\right)\left(\frac{\sin^2 A}{1}\right)$$

$$= \frac{(\sin^2 A - \cos^2 A)(\sin^2 A)}{\sin^2 A}$$

$$= \sin^2 A - \cos^2 A$$

$$\equiv \text{RHS.}$$

$$\cos^2 A + \cot^2 A \cos^2 A \equiv \cot^2 A$$

Eyeballing and Mental Gymnastics

1. $\cot = c/s$
2. c^2, $\cot^2$ suggest $s^2 + c^2 \equiv 1$
3. *rearrange and simplify.*

$$\begin{aligned}
\text{LHS} &= \cos^2 A + \cot^2 A \cos^2 A \\
&= \cos^2 A (1 + \cot^2 A) \\
&= \cos^2 A (\operatorname{cosec}^2 A) \\
&= \cos^2 A \cdot \frac{1}{\sin^2 A} \\
&= \cot^2 A \\
&\equiv \text{RHS.}
\end{aligned}$$

$$(1+\tan A)^2 + (1-\tan A)^2 \equiv 2\sec^2 A$$

Eyeballing and Mental Gymnastics

1. t^2, $\sec^2$ *suggest* $s^2 + c^2 \equiv 1$
2. *rearrange and simplify.*

$$
\begin{aligned}
\text{LHS} &= (1+\tan A)^2 + (1-\tan A)^2 \\
&= (1 + 2\tan A + \tan^2 A) + (1 - 2\tan A + \tan^2 A) \\
&= 2 + 2\tan^2 A \\
&= 2(1 + \tan^2 A) \\
&= 2\sec^2 A \\
&\equiv \text{RHS.}
\end{aligned}
$$

$$\boxed{\operatorname{cosec}^2 A + \sec^2 A \equiv \operatorname{cosec}^2 A \sec^2 A}$$

Eyeballing and Mental Gymnastics

1. cosec^2, $\sec^2$ *suggest* $s^2 + c^2 \equiv 1$
2. $\operatorname{cosec} = 1/\sin$, $\sec = 1/\cos$
3. *rearrange and simplify.*

$$\begin{aligned}
\text{LHS} &= \operatorname{cosec}^2 A + \sec^2 A \\[2mm]
&= \frac{1}{\sin^2 A} + \frac{1}{\cos^2 A} \\[2mm]
&= \frac{\cos^2 A + \sin^2 A}{\sin^2 A \cos^2 A} \\[2mm]
&= \frac{1}{\sin^2 A \cos^2 A} \\[2mm]
&= \operatorname{cosec}^2 A \sec^2 A \\[2mm]
&\equiv \text{RHS.}
\end{aligned}$$

$$2 \sec^2 A - 2 \sec^2 A \sin^2 A - \sin^2 A - \cos^2 A \equiv 1$$

Eyeballing and Mental Gymnastics

1. $\sec^2$, s^2, c^2 *suggest* $s^2 + c^2 \equiv 1$
2. $\sec = 1/\cos$
3. *rearrange and simplify.*

$$\text{LHS} = 2 \sec^2 A - 2 \sec^2 A \sin^2 A - \sin^2 A - \cos^2 A$$

$$= 2 \sec^2 A (1 - \sin^2 A) - (\sin^2 A + \cos^2 A)$$

$$= 2 \frac{1}{\cos^2 A} \cdot (\cos^2 A) - (1)$$

$$= 2 - 1$$

$$= 1$$

$$\equiv \text{RHS}.$$

$$\boxed{\cot^4 A + \cot^2 A \equiv \operatorname{cosec}^4 A - \operatorname{cosec}^2 A}$$

Eyeballing and Mental Gymnastics

1. $\cot = c/s$, $\operatorname{cosec} = 1/\sin$
2. $\cot^2$, cosec^2 *suggest* $s^2 + c^2 \equiv 1$
3. *rearrange and simplify.*

$$\begin{aligned}
\text{LHS} &= \cot^4 A + \cot^2 A \\[2mm]
&= \cot^2 A(\cot^2 A + 1) \\[2mm]
&= \cot^2 A(\operatorname{cosec}^2 A) \\[2mm]
&= (\operatorname{cosec}^2 A - 1)(\operatorname{cosec}^2 A) \\[2mm]
&= \operatorname{cosec}^4 A - \operatorname{cosec}^2 A \\[2mm]
&\equiv \text{RHS.}
\end{aligned}$$

$$\boxed{\frac{\tan A - \cot A}{\tan A + \cot A} \equiv \sin^2 A - \cos^2 A}$$

Eyeballing and Mental Gymnastics

1. $t = s/c$, $\cot = c/s$
2. *common denominator*
3. $s^2 + c^2 \equiv 1$
4. *rearrange and simplify.*

$$\text{LHS} = \frac{\tan A - \cot A}{\tan A + \cot A}$$

$$= \frac{\dfrac{\sin A}{\cos A} - \dfrac{\cos A}{\sin A}}{\dfrac{\sin A}{\cos A} + \dfrac{\cos A}{\sin A}}$$

$$= \frac{\dfrac{\sin^2 A - \cos^2 A}{\cos A \sin A}}{\dfrac{\sin^2 A + \cos^2 A}{\cos A \sin A}}$$

$$= \frac{\sin^2 A - \cos^2 A}{\sin^2 A + \cos^2 A}$$

$$= \sin^2 A - \cos^2 A$$

$$\equiv \text{RHS.}$$

$$\frac{\sec A - \cos A}{\sec A + \cos A} \equiv \frac{\sin^2 A}{1 + \cos^2 A}$$

Eyeballing and Mental Gymnastics

1. $\sec = 1/\cos$
2. *common denominator*
3. $s^2 + c^2 \equiv 1$
4. *rearrange and simplify.*

$$\text{LHS} = \frac{\sec A - \cos A}{\sec A + \cos A}$$

$$= \frac{\dfrac{1}{\cos A} - \cos A}{\dfrac{1}{\cos A} + \cos A}$$

$$= \frac{\dfrac{1 - \cos^2 A}{\cos A}}{\dfrac{1 + \cos^2 A}{\cos A}}$$

$$= \frac{1 - \cos^2 A}{1 + \cos^2 A}$$

$$= \frac{\sin^2 A}{1 + \cos^2 A}$$

$$\equiv \text{RHS.}$$

$$\frac{\sin^3 A + \cos^3 A}{1 - 2\cos^2 A} \equiv \frac{\sec A - \sin A}{\tan A - 1}$$

Eyeballing and Mental Gymnastics

1. $(s^3 + c^3)$ *suggests* $(s+c)(s^2 - sc + c^2)$
2. $(1 - 2\cos^2 A)$ *suggests* $(s^2 - c^2)$ *and* $(s+c)(s-c)$
3. $\sec = 1/\cos, t = s/c$
4. *rearrange and simplify.*

$$\text{LHS} = \frac{\sin^3 A + \cos^3 A}{1 - 2\cos^2 A}$$

$$= \frac{(\sin A + \cos A)(\sin^2 A - \sin A \cos A + \cos^2 A)}{\sin^2 A - \cos^2 A}$$

$$= \frac{(\sin A + \cos A)(1 - \sin A \cos A)}{(\sin A + \cos A)(\sin A - \cos A)}$$

$$= \frac{1 - \sin A \cos A}{\sin A - \cos A}$$

$$= \frac{\sec A - \sin A}{\tan A - 1} \qquad \text{divide both numerator and denominator by } \cos A$$

$$\equiv \text{RHS}.$$

$$\frac{\sec^2 A - \tan^2 A + \tan A}{\sec A} \equiv \sin A + \cos A$$

Eyeballing and Mental Gymnastics

1. $\sec = 1/\cos, t = s/c$
2. $s^2 + c^2 \equiv 1$
3. *rearrange and simplify.*

$$\text{LHS} = \frac{\sec^2 - \tan^2 A + \tan A}{\sec A}$$

$$= \frac{(1 + \tan^2 A) - \tan^2 A + \tan A}{\sec A}$$

$$= \frac{1 + \tan A}{\sec A}$$

$$= \frac{1}{\sec A} + \frac{\sin A}{\cos A} \cdot \frac{\cos A}{1}$$

$$= \cos A + \sin A$$

$$\equiv \text{RHS.}$$

$$1 - \frac{\cos^2 A}{1 + \sin A} \equiv \sin A$$

Eyeballing and Mental Gymnastics

1. $s^2 + c^2 \equiv 1$
2. *rearrange and simplify.*

$$\text{LHS} = 1 - \frac{\cos^2 A}{1 + \sin A}$$

$$= 1 - \frac{(1 - \sin^2 A)}{1 + \sin A}$$

$$= 1 - \frac{(1 + \sin A)(1 - \sin A)}{(1 + \sin A)}$$

$$= 1 - (1 - \sin A)$$

$$= \sin A$$

$$\equiv \text{RHS}.$$

$$(\sec A - \tan A)(\sec A + \tan A) \equiv 1$$

Eyeballing and Mental Gymnastics

1. $(\sec - \tan)(\sec + \tan)$ *suggests* $(a-b)(a+b) = a^2 - b^2$
2. $s^2 + c^2 \equiv 1$
3. *rearrange and simplify.*

$$\text{LHS} = (\sec A - \tan A)(\sec A + \tan A)$$

$$= \sec^2 - \tan^2 A$$

$$= (1 + \tan^2 A) - \tan^2 A$$

$$= 1$$

$$\equiv \text{RHS.}$$

$$\boxed{\sin A + \sin A \cot^2 A \equiv \operatorname{cosec} A}$$

Eyeballing and Mental Gymnastics

1. $\cot^2$ *suggests* $s^2 + c^2 \equiv 1$
2. $\operatorname{cosec} = 1/\sin$
3. *rearrange and simplify.*

$$\begin{aligned}
\text{LHS} &= \sin A + \sin A \cot^2 A \\
&= \sin A (1 + \cot^2 A) \\
&= \sin A \operatorname{cosec}^2 A \\
&= \sin A \cdot \frac{1}{\sin^2 A} \\
&= \frac{1}{\sin A} \\
&= \operatorname{cosec} A \\
&\equiv \text{RHS.}
\end{aligned}$$

$$(1 - \sin A)(\sec A + \tan A) \equiv \cos A$$

Eyeballing and Mental Gymnastics

1. $\sec = 1/\cos$, $t = s/c$
2. *common denominator*
3. *rearrange and simplify.*

$$\text{LHS} = (1 - \sin A)(\sec A + \tan A)$$

$$= (1 - \sin A)\left(\frac{1}{\cos A} + \frac{\sin A}{\cos A}\right)$$

$$= (1 - \sin A)\left(\frac{1 + \sin A}{\cos A}\right)$$

$$= \frac{1 - \sin^2 A}{\cos A}$$

$$= \frac{\cos^2 A}{\cos A}$$

$$= \cos A$$

$$\equiv \text{RHS.}$$

$$\boxed{\tan A \sin A \equiv \sec A - \cos A}$$

Eyeballing and Mental Gymnastics

1. $t = s/c$, $\sec = 1/\cos$
2. *rearrange and simplify.*

$$
\begin{aligned}
\text{LHS} &= \tan A \sin A \\
&= \frac{\sin A}{\cos A} \cdot \sin A \\
&= \frac{\sin^2 A}{\cos A} \\
&= \frac{(1 - \cos^2 A)}{\cos A} \\
&= \frac{1}{\cos A} - \cos A \\
&= \sec A - \cos A \\
&\equiv \text{RHS.}
\end{aligned}
$$

$$\boxed{\tan A(\sin A + \cot A \cos A) \equiv \sec A}$$

Eyeballing and Mental Gymnastics

1. $t = s/c$, $\cot = c/s$, $\operatorname{cosec} = 1/\sin$
2. *simplify.*

$$\text{LHS} = \tan A(\sin A + \cot A \cos A)$$

$$= \frac{\sin A}{\cos A}\left(\sin A + \frac{\cos A}{\sin A}\cdot \cos A\right)$$

$$= \frac{\sin^2 A}{\cos A} + \cos A$$

$$= \frac{\sin^2 A + \cos^2 A}{\cos A}$$

$$= \frac{1}{\cos A}$$

$$= \sec A$$

$$\equiv \text{RHS.}$$

$$\boxed{(1 - \cos A)(1 + \sec A) \equiv \sin A \tan A}$$

Eyeballing and Mental Gymnastics

1. $\sec = 1/\cos$, $t = s/c$
2. *rearrange and simplify.*

$$\text{LHS} = (1 - \cos A)(1 + \sec A)$$

$$= (1 - \cos A)\left(1 + \frac{1}{\cos A}\right)$$

$$= (1 - \cos A)\left(\frac{1 + \cos A}{\cos A}\right)$$

$$= \frac{1 - \cos^2 A}{\cos A}$$

$$= \frac{\sin^2 A}{\cos A}$$

$$= \sin A \tan A$$

$$\equiv \text{RHS.}$$

$$(1 - \sin A)(1 + \operatorname{cosec} A) \equiv \cos A \cot A$$

Eyeballing and Mental Gymnastics

1. $\operatorname{cosec} = 1/\sin$, $\cot = c/s$
2. *rearrange and simplify.*

$$\text{LHS} = (1 - \sin A)(1 + \operatorname{cosec} A)$$

$$= 1 - \sin A + \operatorname{cosec} A - \sin A \cdot \operatorname{cosec} A$$

$$= \operatorname{cosec} A - \sin A \qquad\qquad \left| \begin{array}{l} \sin A \cdot \operatorname{cosec} A \\ \quad = 1 \end{array} \right.$$

$$= \frac{1}{\sin A} - \sin A$$

$$= \frac{1 - \sin^2 A}{\sin A}$$

$$= \frac{\cos^2 A}{\sin A}$$

$$= \cos A \cot A$$

$$\equiv \text{RHS.}$$

$$2\sec^2 A - 1 \equiv 1 + 2\tan^2 A$$

Eyeballing and Mental Gymnastics

1. $\sec^2$, t^2 *suggest* $s^2 + c^2 \equiv 1$
2. *rearrange and simplify.*

$$\text{LHS} = 2\sec^2 A - 1$$

$$= 2(1 + \tan^2 A) - 1$$

$$= 2 + 2\tan^2 A - 1$$

$$= 1 + 2\tan^2 A$$

$$\equiv \text{RHS}.$$

$$(\sin A - \cos B)^2 + (\cos B + \sin A)(\cos B - \sin A) = -2\cos B(\sin A - \cos B)$$

Eyeballing and Mental Gymnastics

1. *Although the identity looks lengthy and complicated, closer inspection shows that the term* $(\sin A - \cos B)$ *is common on both sides of the equation.*
2. *Rearrange and simplify.*

$$\begin{aligned}
\text{LHS} &= (\sin A - \cos B)^2 + (\cos B + \sin A)(\cos B - \sin A) \\
&= (\sin A - \cos B)^2 - (\sin A - \cos B)(\cos B + \sin A) \\
&= (\sin A - \cos B)((\sin A - \cos B) - (\cos B - \sin A)) \\
&= (\sin A - \cos B)(-2\cos B) \\
&= -2\cos B(\sin A - \cos B) \\
&\equiv \text{RHS}.
\end{aligned}$$

$$\boxed{\frac{\sec A}{1 + \sec A} \equiv \frac{1 - \cos A}{\sin^2 A}}$$

Eyeballing and Mental Gymnastics

1. $\sec = 1/\cos$
2. *common denominator*
3. $s^2 + c^2 \equiv 1$
4. *rearrange and simplify.*

$$\text{LHS} = \frac{\sec A}{1 + \sec A}$$

$$= \frac{1}{\cos A} \cdot \frac{1}{1 + \dfrac{1}{\cos A}}$$

$$= \frac{1}{\cos A} \cdot \frac{\cos A}{(\cos A + 1)}$$

$$= \frac{1}{\cos A + 1}$$

$$= \left(\frac{1 - \cos A}{1 - \cos A}\right) \cdot \frac{1}{(1 + \cos A)}$$

$$= \frac{1 - \cos A}{1 - \cos^2 A}$$

$$= \frac{1 - \cos A}{\sin^2 A}$$

$$\equiv \text{RHS}.$$

$$\boxed{\frac{1 + \sec A}{\tan A + \sin A} \equiv \operatorname{cosec} A}$$

Eyeballing and Mental Gymnastics

1. $\sec = 1/\cos$, $t = s/c$, $\operatorname{cosec} = 1/\sin$
2. *rearrange and simplify.*

$$\text{LHS} = \frac{1 + \sec A}{\tan A + \sin A}$$

$$= \left(1 + \frac{1}{\cos A}\right)\left(\frac{1}{\frac{\sin A}{\cos A} + \sin A}\right)$$

$$= \left(\frac{\cos A + 1}{\cos A}\right)\left(\frac{\cos A}{\sin A + \sin A \cos A}\right)$$

$$= \frac{\cos A + 1}{\sin A(1 + \cos A)}$$

$$= \frac{1}{\sin A}$$

$$= \operatorname{cosec} A$$

$$\equiv \text{RHS.}$$

$$\frac{\cos A}{\operatorname{cosec} A - 1} + \frac{\cos A}{\operatorname{cosec} + 1} \equiv 2\tan A$$

Eyeballing and Mental Gymnastics

1. *common denominator*
2. $t = s/c$, $\operatorname{cosec} = 1/\sin$
3. *rearrange and simplify.*

$$\text{LHS} = \frac{\cos A}{\operatorname{cosec} A - 1} + \frac{\cos A}{\operatorname{cosec} A + 1}$$

$$= \frac{\cos A(\operatorname{cosec} A + 1) + \cos A(\operatorname{cosec} A - 1)}{(\operatorname{cosec} A - 1)(\operatorname{cosec} A + 1)}$$

$$= \frac{(\cot A + \cos A) + (\cot A - \cos A)}{\operatorname{cosec}^2 A - 1}$$

$$= \frac{2\cot A}{\cot^2 A} \qquad |\operatorname{cosec}^2 - 1 = \cot^2$$

$$= \frac{2}{\cot A}$$

$$= 2\tan A$$

$$\equiv \text{RHS.}$$

$$\frac{\cos A}{1 - \tan A} + \frac{\sin A}{1 - \cot A} \equiv \sin A + \cos A$$

Eyeballing and Mental Gymnastics

1. $t = s/c$, $\cot = c/s$
2. *common denominator*
3. *rearrange and simplify.*

$$\text{LHS} = \frac{\cos A}{1 - \tan A} + \frac{\sin A}{1 - \cot A}$$

$$= \frac{\cos A}{\left(1 - \dfrac{\sin A}{\cos A}\right)} + \frac{\sin A}{\left(1 - \dfrac{\cos A}{\sin A}\right)}$$

$$= \cos A \cdot \frac{\cos A}{(\cos A - \sin A)} + \sin A \cdot \frac{\sin A}{(\sin A - \cos A)}$$

$$= \frac{\cos^2 A - \sin^2 A}{\cos A - \sin A}$$

$$= \frac{(\cos A + \sin A)(\cos A - \sin A)}{(\cos A - \sin A)}$$

$$= \cos A + \sin A$$

$$\equiv \text{RHS.}$$

$$\boxed{\frac{\sec A - \cos A}{\tan A} \equiv \sin A}$$

Eyeballing and Mental Gymnastics

1. $\sec = 1/\cos$, $t = s/c$
2. *simplify.*

$$\text{LHS} = \frac{\sec A - \cos A}{\tan A}$$

$$= \left(\frac{1}{\cos A} - \cos A\right)\left(\frac{\cos A}{\sin A}\right)$$

$$= \left(\frac{1 - \cos^2 A}{\cos A}\right)\left(\frac{\cos A}{\sin A}\right)$$

$$= \frac{1 - \cos^2 A}{\sin A}$$

$$= \frac{\sin^2 A}{\sin A}$$

$$= \sin A$$

$$\equiv \text{RHS.}$$

$$\boxed{\frac{\tan A + \sec A - 1}{\tan A - \sec A + 1} \equiv \tan A + \sec A}$$

Eyeballing and Mental Gymnastics

1. $s^2 + c^2 \equiv 1$
    $\tan^2 + 1 \equiv \sec^2$
    $\qquad 1 \equiv \sec^2 - \tan^2$
2. *rearrange and simplify.*

$$
\begin{aligned}
\text{LHS} &= \frac{\tan A + \sec A - 1}{\tan A - \sec A + 1} \\[2mm]
&= \frac{\tan A + \sec A - (\sec^2 A - \tan^2 A)}{\tan A - \sec A + 1} \\[2mm]
&= \frac{(\tan A + \sec A) - (\sec A + \tan A)(\sec A - \tan A)}{\tan A - \sec A + 1} \\[2mm]
&= \frac{(\sec A + \tan A)(1 - \sec A + \tan A)}{\tan A - \sec A + 1} \\[2mm]
&= \frac{(\sec A + \tan A)(\tan A - \sec A + 1)}{(\tan A - \sec A + 1)} \\[2mm]
&= \tan A + \sec A \\[2mm]
&\equiv \text{RHS.}
\end{aligned}
$$

$$\boxed{\frac{\sin A - \cos A + 1}{\sin A + \cos A - 1} \equiv \frac{\sin A + 1}{\cos A}}$$

Eyeballing and Mental Gymnastics

1. $s^2 + c^2 \equiv 1$
2. *multiple terms in denominator; may have to use $s^2 + c^2 = 1$ to arrive at common factor with numerator*
3. *rearrange and simplify.*

$$\text{LHS} = \frac{\sin A - \cos A + 1}{\sin A + \cos A - 1}$$

$$= \frac{\sin A - \cos A + 1}{\sin A + \cos A - 1} \cdot \left(\frac{\sin A + 1}{\sin A + 1} \right)$$

$$= \frac{(\sin A - \cos A + 1)(\sin A + 1)}{(\sin^2 A + \sin A \cos A - \sin A) + (\sin A + \cos A - 1)}$$

$$= \frac{(\sin A - \cos A + 1)(\sin A + 1)}{-\cos^2 A + \sin A \cos A + \cos A}$$

$$= \frac{(\sin A - \cos A + 1)(\sin A + 1)}{\cos A(-\cos A + \sin A + 1)}$$

$$= \frac{\sin A + 1}{\cos A}$$

$$\equiv \text{RHS.}$$

$$\boxed{\frac{\sec A + \tan A}{\cot A + \cos A} \equiv \tan A \sec A}$$

Eyeballing and Mental Gymnastics

1. $\sec = 1/\cos$, $t = s/c$, $\cot = c/s$
2. *common denominator*
3. *rearrange and simplify.*

$$\text{LHS} = \frac{\sec A + \tan A}{\cot A + \cos A}$$

$$= \frac{\dfrac{1}{\cos A} + \dfrac{\sin A}{\cos A}}{\dfrac{\cos A}{\sin A} + \cos A}$$

$$= \frac{\dfrac{1 + \sin A}{\cos A}}{\dfrac{\cos A + \cos A \sin A}{\sin A}}$$

$$= \frac{(1 + \sin A)}{\cos A} \cdot \frac{\sin A}{\cos A(1 + \sin A)}$$

$$= \frac{\sin A}{\cos A} \cdot \frac{1}{\cos A}$$

$$= \tan A \sec A$$

$$\equiv \text{RHS}.$$

$$\frac{1 + \sin A + \cos A}{1 + \sin A - \cos A} \equiv \frac{1 + \cos A}{\sin A}$$

Eyeballing and Mental Gymnastics

1. *multiple term in denominator; may have to find common factor for cancellation.*
2. $s^2 + c^2 \equiv 1$
3. *rearrange and simplify.*

$$
\begin{aligned}
\text{LHS} &= \frac{1 + \sin A + \cos A}{1 + \sin A - \cos A} \\[2mm]
&= \frac{1 + \sin A + \cos A}{1 + \sin A - \cos A} \cdot \left(\frac{1 + \cos A}{1 + \cos A} \right) \\[2mm]
&= \frac{(1 + \sin A + \cos A)(1 + \cos A)}{(1 + \sin A - \cos A) + (\cos A + \sin A \cos A - \cos^2 A)} \\[2mm]
&= \frac{(1 + \sin A + \cos A)(1 + \cos A)}{\sin^2 A + \sin A + \sin A \cos A} \\[2mm]
&= \frac{(1 + \sin A + \cos A)(1 + \cos A)}{\sin A (\sin A + 1 + \cos A)} \\[2mm]
&= \frac{1 + \cos A}{\sin A} \\[2mm]
&\equiv \text{RHS}.
\end{aligned}
$$

$$\boxed{\frac{1 + \cos A}{\sin A} + \frac{\sin A}{1 + \cos A} \equiv 2 \operatorname{cosec} A}$$

Eyeballing and Mental Gymnastics

1. *common denominator*
2. *cosec = 1/sin*
3. *rearrange and simplify.*

$$\mathrm{LHS} = \frac{1 + \cos A}{\sin A} + \frac{\sin A}{1 + \cos A}$$

$$= \frac{(1 + \cos A)^2 + \sin^2 A}{\sin A(1 + \cos A)}$$

$$= \frac{1 + 2\cos A + \cos^2 A + \sin^2 A}{\sin A(1 + \cos A)}$$

$$= \frac{1 + 2\cos A + 1}{\sin A(1 + \cos A)}$$

$$= \frac{2(1 + \cos A)}{\sin A(1 + \cos A)}$$

$$= \frac{2}{\sin A}$$

$$= 2 \operatorname{cosec} A$$

$$\equiv \mathrm{RHS}.$$

$$\boxed{\frac{\sec A}{\sin A} - \frac{\sin A}{\cos A} \equiv \cot A}$$

Eyeballing and Mental Gymnastics

1. $\sec = 1/\cos$, $\cot = c/s$
2. *common denominator*
3. *rearrange and simplify.*

$$\text{LHS} = \frac{\sec A}{\sin A} - \frac{\sin A}{\cos A}$$

$$= \frac{1}{\cos A} \cdot \frac{1}{\sin A} - \frac{\sin A}{\cos A}$$

$$= \frac{1 - \sin^2 A}{\cos A \sin A}$$

$$= \frac{\cos^2 A}{\cos A \sin A}$$

$$= \frac{\cos A}{\sin A}$$

$$= \cot A$$

$$= \text{RHS.}$$

$$\boxed{\frac{1-\sin A}{\cos A} + \frac{\cos A}{1-\sin A} \equiv 2\sec A}$$

Eyeballing and Mental Gymnastics

1. $\sec = 1/\cos$
2. *common denominator*
3. $s^2 + c^2 \equiv 1$
4. *rearrange and simplify.*

$$\text{LHS} = \frac{1-\sin A}{\cos A} + \frac{\cos A}{1-\sin A}$$

$$= \frac{(1-\sin A)^2 + \cos^2 A}{\cos A(1-\sin A)}$$

$$= \frac{(1-\sin A)^2 + (1-\sin^2 A)}{\cos A(1-\sin A)}$$

$$= \frac{(1-\sin A)^2 + (1-\sin A)(1+\sin A)}{\cos A(1-\sin A)}$$

$$= \frac{(1-\sin A)(1-\sin A + 1+\sin A)}{\cos A(1-\sin A)}$$

$$= \frac{2}{\cos A}$$

$$= 2\sec A$$

$$\equiv \text{RHS}.$$

$$\boxed{\frac{\operatorname{cosec} A}{1 + \operatorname{cosec} A} \equiv \frac{1 - \sin A}{\cos^2 A}}$$

Eyeballing and Mental Gymnastics

1. $\operatorname{cosec} = 1/\sin$
2. *common denominator*
3. $s^2 + c^2 \equiv 1$
4. *rearrange and simplify.*

$$\text{LHS} = \frac{\operatorname{cosec} A}{1 + \operatorname{cosec} A}$$

$$= \frac{1}{\sin A} \left(\frac{1}{1 + \dfrac{1}{\sin A}} \right)$$

$$= \frac{1}{\sin A} \left(\frac{\sin A}{\sin A + 1} \right)$$

$$= \frac{1}{1 + \sin A}$$

$$= \frac{1}{(1 + \sin A)} \left(\frac{1 - \sin A}{1 - \sin A} \right)$$

$$= \frac{1 - \sin A}{1 - \sin^2 A}$$

$$= \frac{1 - \sin A}{\cos^2 A}$$

$$\equiv \text{RHS}.$$

$$\frac{\cos A}{1+\sin A} + \frac{1+\sin A}{\cos A} \equiv 2\sec A$$

Eyeballing and Mental Gymnastics

1. $\sec = 1/\cos$
2. *common denominator*
3. $s^2 + c^2 \equiv 1$
4. *rearrange and simplify.*

$$\text{LHS} = \frac{\cos A}{1+\sin A} + \frac{1+\sin A}{\cos A}$$

$$= \frac{\cos^2 A + (1+\sin A)^2}{\cos A(1+\sin A)}$$

$$= \frac{(1-\sin^2 A) + (1+\sin A)^2}{\cos A(1+\sin A)}$$

$$= \frac{(1+\sin A)(1-\sin A) + (1+\sin A)^2}{\cos A(1+\sin A)}$$

$$= \frac{(1+\sin A)(1-\sin A + 1 + \sin A)}{\cos A(1+\sin A)}$$

$$= \frac{2}{\cos A}$$

$$= 2\sec A$$

$$\equiv \text{RHS}.$$

$$\boxed{\frac{1+\cos A + \sin A}{1+\cos A - \sin A} \equiv \sec A + \tan A}$$

Eyeballing and Mental Gymnastics

1. *multiple terms in denominator suggest use of common factor for cancellation*
2. *sec + tan in RHS suggest $1/\cos + \sin/\cos$ which gives $(1+\sin)/\cos$*
3. *$s^2 + c^2 \equiv 1$*
4. *rearrange and simplify.*

$$\text{LHS} = \frac{1+\cos A + \sin A}{1+\cos A - \sin A}$$

$$= \frac{1+\cos A + \sin A}{1+\cos A - \sin A} \cdot \left(\frac{1+\sin A}{1+\sin A}\right)$$

$$= \frac{(1+\cos A + \sin A)(1+\sin A)}{1+\cos A - \sin A + \sin A + \sin A \cos A - \sin^2 A}$$

$$= \frac{(1+\cos A + \sin A)(1+\sin A)}{\cos^2 A + \cos A + \sin A \cos A}$$

$$= \frac{(1+\cos A + \sin A)(1+\sin A)}{\cos A(\cos A + 1 + \sin A)}$$

$$= \frac{1+\sin A}{\cos A}$$

$$= \sec A + \tan A$$

$$\equiv \text{RHS.}$$

$$\boxed{\sec A - \tan A \equiv \frac{\cos A}{1 + \sin A}}$$

Eyeballing and Mental Gymnastics

1. $\sec = 1/\cos, t = s/c$
2. $s^2 + c^2 \equiv 1$
3. *rearrange and simplify.*

$$\mathrm{LHS} = \sec A - \tan A$$

$$= \frac{1}{\cos A} - \frac{\sin A}{\cos A}$$

$$= \frac{1 - \sin A}{\cos A}$$

$$= \frac{1 - \sin A}{\cos A} \left(\frac{1 + \sin A}{1 + \sin A} \right)$$

$$= \frac{1 - \sin^2 A}{\cos A (1 + \sin A)}$$

$$= \frac{\cos^2 A}{\cos A (1 + \sin A)}$$

$$= \frac{\cos A}{1 + \sin A}$$

$$\equiv \mathrm{RHS}.$$

$$\boxed{\operatorname{cosec} A - \cot A \equiv \frac{\sin A}{1 + \cos A}}$$

Eyeballing and Mental Gymnastics

1. $\operatorname{cosec} = 1/\sin$, $\cot = c/s$
2. $s^2 + c^2 \equiv 1$
3. *rearrange and simplify.*

$$\text{LHS} = \operatorname{cosec} A - \cot A$$

$$= \frac{1}{\sin A} - \frac{\cos A}{\sin A}$$

$$= \frac{1 - \cos A}{\sin A}$$

$$= \frac{1 - \cos A}{\sin A} \left(\frac{1 + \cos A}{1 + \cos A} \right)$$

$$= \frac{1 - \cos^2 A}{\sin A (1 + \cos A)}$$

$$= \frac{\sin^2 A}{\sin A (1 + \cos A)}$$

$$= \frac{\sin A}{1 + \cos A}$$

$$\equiv \text{RHS.}$$

$$\frac{\cot A}{1 - \tan A} + \frac{\tan A}{1 - \cot A} \equiv 1 + \sec A \operatorname{cosec} A$$

Eyeballing and Mental Gymnastics

1. $\cot = c/s, t = s/c, \sec = 1/\cos, \operatorname{cosec} = 1/\sin$
2. *rearrange and simplify.*

$$\text{LHS} = \frac{\cot A}{1 - \tan A} + \frac{\tan A}{1 - \cot A}$$

$$= \frac{\cos A}{\sin A} \frac{1}{\left(1 - \frac{\sin A}{\cos A}\right)} + \frac{\sin A}{\cos A} \cdot \frac{1}{\left(1 - \frac{\cos A}{\sin A}\right)}$$

$$= \frac{\cos A}{\sin A} \cdot \frac{\cos A}{\cos A - \sin A} + \frac{\sin A}{\cos A} \frac{\sin A}{(\sin A - \cos A)}$$

$$= \frac{\cos^2 A}{\sin A(\cos A - \sin A)} - \frac{\sin^2 A}{\cos A(\cos A - \sin A)}$$

$$= \frac{\cos^3 A - \sin^3 A}{(\cos A - \sin A)(\sin A \cos A)}$$

$$= \frac{(\cos A - \sin A)(\cos^2 A + \cos A \sin A + \sin^2 A)}{(\cos A - \sin A)(\sin A \cos A)}$$

$$= \frac{(1 + \cos A \sin A)}{\sin A \cos A}$$

$$= \operatorname{cosec} A \sec A + 1$$

$$\equiv \text{RHS.}$$

$$\frac{\sin A}{1 - \cos A} + \frac{1 - \cos A}{\sin A} \equiv 2 \operatorname{cosec} A$$

Eyeballing and Mental Gymnastics

1. $\operatorname{cosec} = 1/\sin$
2. *common denominator*
3. *rearrange and simplify.*

$$\text{LHS} = \frac{\sin A}{1 - \cos A} + \frac{1 - \cos A}{\sin A}$$

$$= \frac{\sin^2 A + (1 - \cos A)^2}{\sin A (1 - \cos A)}$$

$$= \frac{\sin^2 A + (1 - 2 \cos A + \cos^2 A)}{\sin A (1 - \cos A)}$$

$$= \frac{\sin^2 A + \cos^2 A + 1 - 2 \cos A}{\sin A (1 - \cos A)}$$

$$= \frac{2 - 2 \cos A}{\sin A (1 - \cos A)}$$

$$= \frac{2(1 - \cos A)}{\sin A (1 - \cos A)}$$

$$= \frac{2}{\sin A}$$

$$= 2 \operatorname{cosec} A$$

$$\equiv \text{RHS.}$$

$$\boxed{\frac{\tan A}{\sec A - 1} \equiv \frac{\sec A + 1}{\tan A}}$$

Eyeballing and Mental Gymnastics

1. $t = s/c,\ \sec = 1/\cos c$
2. $s^2 + c^2 \equiv 1$
3. *common denominator*
4. *rearrange and simplify.*

$$\text{LHS} = \frac{\tan A}{\sec A - 1}$$

$$= \frac{\sin A}{\cos A}\left(\frac{1}{\dfrac{1}{\cos A} - 1}\right)$$

$$= \frac{\sin A}{\cos A}\left(\frac{\cos A}{1 - \cos A}\right)$$

$$= \frac{\sin A}{1 - \cos A}$$

$$= \frac{\sin A}{(1 - \cos A)}\left(\frac{1 + \cos A}{1 + \cos A}\right)$$

$$= \frac{\sin A + \sin A \cos A}{1 - \cos^2 A}$$

$$= \frac{\sin A + \sin A \cos A}{\sin^2 A}$$

$$= \frac{1 + \cos A}{\sin A}$$

$$= \frac{\sec A + 1}{\tan A}$$

$$\equiv \text{RHS.}$$

$$\tan A + \frac{\cos A}{1 + \sin A} \equiv \sec A$$

Eyeballing and Mental Gymnastics

1. $t = s/c$, $\sec = 1/\cos$
2. *common denominator*
3. *rearrange and simplify.*

$$\text{LHS} = \tan A + \frac{\cos A}{1 + \sin A}$$

$$= \frac{\sin A}{\cos A} + \frac{\cos A}{1 + \sin A}$$

$$= \frac{\sin A(1 + \sin A) + \cos^2 A}{\cos A(1 + \sin A)}$$

$$= \frac{\sin A + \sin^2 A + \cos^2 A}{\cos A(1 + \sin A)}$$

$$= \frac{(1 + \sin A)}{\cos A(1 + \sin A)}$$

$$= \frac{1}{\cos A}$$

$$= \sec A$$

$$\equiv \text{RHS.}$$

$$\frac{\cot A}{1 - \tan A} + \frac{\tan A}{1 - \cot A} \equiv 1 + \tan A + \cot A$$

Eyeballing and Mental Gymnastics

1. $\cot = c/s, t = s/c$
2. *common denominator*
3. $s^2 + c^2 \equiv 1$
4. *rearrange and simplify.*

$$\text{LHS} = \frac{\cot A}{1 - \tan A} + \frac{\tan A}{1 - \cot A}$$

$$= \frac{\dfrac{\cos A}{\sin A}}{1 - \dfrac{\sin A}{\cos A}} + \frac{\dfrac{\sin A}{\cos A}}{1 - \dfrac{\cos A}{\sin A}}$$

$$= \frac{\cos A}{\sin A} \cdot \frac{\cos A}{(\cos A - \sin A)} + \frac{\sin A}{\cos A} \cdot \frac{\sin A}{(\sin A - \cos A)}$$

$$= \frac{\cos^3 A - \sin^3 A}{\sin A \cos A (\cos A - \sin A)}$$

$$= \frac{(\cos A - \sin A)(\cos^2 A + \sin A \cos A + \sin^2 A)}{(\cos A - \sin A)(\sin A \cos A)}$$

$$= \cot A + 1 + \tan A$$

$$= 1 + \tan A + \cot A$$

$$\equiv \text{RHS.}$$

$$\boxed{\frac{\operatorname{cosec} A - 1}{\cot A} \equiv \frac{\cot A}{\operatorname{cosec} A + 1}}$$

Eyeballing and Mental Gymnastics

1. $\operatorname{cosec} = 1/\sin,\ \cot = c/s$
2. $s^2 + c^2 \equiv 1$
3. *rearrange and simplify.*

$$\text{LHS} = \frac{\operatorname{cosec} A - 1}{\cot A}$$

$$= \frac{\dfrac{1}{\sin A} - 1}{\dfrac{\cos A}{\sin A}}$$

$$= \frac{1 - \sin A}{\sin A} \cdot \frac{\sin A}{\cos A}$$

$$= \frac{1 - \sin A}{\cos A}$$

$$= \frac{1 - \sin A}{\cos A} \left(\frac{1 + \sin A}{1 + \sin A} \right)$$

$$= \frac{1 - \sin^2 A}{\cos A (1 + \sin A)}$$

$$= \frac{\cos^2 A}{\cos A (1 + \sin A)}$$

$$= \frac{\cos A}{1 + \sin A}$$

$$= \frac{\cot A}{\operatorname{cosec} A + 1}$$

$$\equiv \text{RHS.}$$

| divide both
| numerator
| and denominator
| by sin

$$\boxed{\frac{\operatorname{cosec} A}{1 - \cos A} \equiv \frac{1 + \cos A}{\sin^3 A}}$$

Eyeballing and Mental Gymnastics

1. $\operatorname{cosec} = 1/\sin$
2. $\sin^3$ *suggests* $s \cdot s^2$, $s^2 + c^2 = 1$
3. *rearrange and simplify.*

$$
\begin{aligned}
\text{LHS} &= \frac{\operatorname{cosec} A}{1 - \cos A} \\[2mm]
&= \frac{1}{\sin A} \cdot \frac{1}{(1 - \cos A)} \\[2mm]
&= \frac{1}{\sin A} \cdot \frac{1}{(1 - \cos A)} \cdot \left(\frac{1 + \cos A}{1 + \cos A} \right) \\[2mm]
&= \frac{1 + \cos A}{\sin A (1 - \cos^2 A)} \\[2mm]
&= \frac{1 + \cos A}{\sin A (\sin^2 A)} \\[2mm]
&= \frac{1 + \cos A}{\sin^3 A} \\[2mm]
&\equiv \text{RHS.}
\end{aligned}
$$

$$1 - \frac{\sin^2 A}{1 + \cos A} \equiv \cos A$$

Eyeballing and Mental Gymnastics

1. *s^2 suggests $s^2 + c^2 \equiv 1$*
2. *rearrange and simplify.*

$$\text{LHS} = 1 - \frac{\sin^2 A}{1 + \cos A}$$

$$= \frac{(1 + \cos A) - \sin^2 A}{1 + \cos A}$$

$$= \frac{\cos^2 A + \cos A}{1 + \cos A}$$

$$= \frac{\cos A (\cos A + 1)}{1 + \cos A}$$

$$= \cos A$$

$$\equiv \text{RHS.}$$

$$\frac{5 - 10\cos^2 A}{\sin A - \cos A} \equiv 5(\sin A + \cos A)$$

Eyeballing and Mental Gymnastics

1. c^2 suggests $s^2 + c^2 \equiv 1$
2. *rearrange and simplify.*

$$\begin{aligned}
\text{LHS} &= \frac{5 - 10\cos^2 A}{\sin A - \cos A} \\[2mm]
&= \frac{5(1 - 2\cos^2 A)}{\sin A - \cos A} \\[2mm]
&= \frac{5(\cos^2 A + \sin^2 A - 2\cos^2 A)}{\sin A - \cos A} \\[2mm]
&= \frac{5(\sin^2 A - \cos^2 A)}{\sin A - \cos A} \\[2mm]
&= \frac{5(\sin A + \cos A)(\sin A - \cos A)}{(\sin A - \cos A)} \\[2mm]
&= 5(\sin A + \cos A) \\[2mm]
&\equiv \text{RHS.}
\end{aligned}$$

$$\boxed{\dfrac{1+\sin A}{1-\sin A} \equiv (\sec A + \tan A)^2}$$

Eyeballing and Mental Gymnastics

1. $(\sec + \tan)^2$ *suggests* $s^2 + c^2 \equiv 1$
2. $\sec = 1/\cos,\ t = s/c$
3. *rearrange and simplify.*

$$\text{LHS} = \frac{1+\sin A}{1-\sin A}$$

$$= \frac{1+\sin A}{1-\sin A}\left(\frac{1+\sin A}{1+\sin A}\right)$$

$$= \frac{(1+\sin A)^2}{1-\sin^2 A}$$

$$= \frac{(1+\sin A)^2}{\cos^2 A}$$

$$= \left(\frac{1+\sin A}{\cos A}\right)^2$$

$$= (\sec A + \tan A)^2$$

$$\equiv \text{RHS}.$$

$$\boxed{(\sec A - \tan A)^2 \equiv \frac{1 - \sin A}{1 + \sin A}}$$

Eyeballing and Mental Gymnastics

1. $\sec = 1/\cos$, $t = s/c$
2. $(\ \ )^2$ suggests $s^2 + c^2 \equiv 1$
3. *common denominator*
4. *rearrange and simplify.*

$$\text{LHS} = (\sec A - \tan A)^2$$

$$= \left(\frac{1}{\cos A} - \frac{\sin A}{\cos A} \right)^2$$

$$= \frac{(1 - \sin A)^2}{\cos^2 A}$$

$$= \frac{(1 - \sin A)^2}{(1 - \sin^2 A)}$$

$$= \frac{(1 - \sin A)(1 - \sin A)}{(1 - \sin A)(1 + \sin A)}$$

$$= \frac{1 - \sin A}{1 + \sin A}$$

$$\equiv \text{RHS.}$$

$$\boxed{\frac{1-\cos A}{1+\cos A} \equiv (\operatorname{cosec} A - \cot A)^2}$$

Eyeballing and Mental Gymnastics

1. $\operatorname{cosec} = 1/\sin$, $\cot = c/s$
2. $(\;\;)^2$ *suggests* $s^2 + c^2 \equiv 1$
3. *rearrange and simplify.*

$$\text{LHS} = \frac{1-\cos A}{1+\cos A}$$

$$= \frac{(1-\cos A)}{(1+\cos A)} \cdot \left(\frac{1-\cos A}{1-\cos A}\right)$$

$$= \frac{(1-\cos A)^2}{(1-\cos^2 A)}$$

$$= \frac{(1-\cos A)^2}{\sin^2 A}$$

$$= \left(\frac{1}{\sin A} - \frac{\cos A}{\sin A}\right)^2$$

$$= (\operatorname{cosec} A - \cot A)^2$$

$$\equiv \text{RHS}.$$

$$1 + \frac{\cos^2 A}{\sin A - 1} \equiv -\sin A$$

Eyeballing and Mental Gymnastics

1. c^2 *suggests* $s^2 + c^2 \equiv 1$
2. *rearrange and simplify.*

$$\text{LHS} = 1 + \frac{\cos^2 A}{\sin A - 1}$$

$$= \frac{(\sin A - 1) + \cos^2 A}{\sin A - 1}$$

$$= \frac{(\sin A - \sin^2 A - \cos^2 A) + \cos^2 A}{\sin A - 1}$$

$$= \frac{\sin A - \sin^2 A}{\sin A - 1}$$

$$= \frac{\sin A(1 - \sin A)}{\sin A - 1}$$

$$= -\sin A$$

$$\equiv \text{RHS.}$$

$$(1 + \cot A)^2 + (1 - \cot A)^2 \equiv \frac{2}{\sin^2 A}$$

Eyeballing and Mental Gymnastics

1. $(\quad)^2$, s^2 *suggest* $s^2 + c^2 \equiv 1$
2. $\cot = c/s$
3. *rearrange and simplify.*

$$\text{LHS} = (1 + \cot A)^2 + (1 - \cot A)^2$$

$$= (1 + 2\cot A + \cot^2 A) + (1 - 2\cot A + \cot^2 A)$$

$$= 2 + 2\cot^2 A$$

$$= 2(1 + \cot^2 A)$$

$$= 2\operatorname{cosec}^2 A$$

$$= \frac{2}{\sin^2 A}$$

$$\equiv \text{RHS.}$$

$$\frac{\sin A \cos B}{\cos A \sin B}(\cot A \cot B) + 1 \equiv \frac{1}{\sin^2 B}$$

Eyeballing and Mental Gymnastics

1. $\cot = c/s$
2. s^2 *suggests* $s^2 + c^2 \equiv 1$
3. *rearrange and simplify.*

$$\begin{aligned}
\text{LHS} &= \frac{\sin A \cos B}{\cos A \sin B}(\cot A \cot B) + 1 \\[2mm]
&= \frac{\sin A \cos B}{\cos A \sin B} \cdot \frac{\cos A}{\sin A} \cdot \frac{\cos B}{\sin B} + 1 \\[2mm]
&= \frac{\cos^2 B}{\sin^2 B} + 1 \\[2mm]
&= \frac{\cos^2 B + \sin^2 B}{\sin^2 B} \\[2mm]
&= \frac{1}{\sin^2 B} \\[2mm]
&\equiv \text{RHS}.
\end{aligned}$$

$$\boxed{\operatorname{cosec} A(\operatorname{cosec} A - \sin A) + \frac{\sin A - \cos A}{\sin A} + \cot A \equiv \operatorname{cosec}^2 A}$$

Eyeballing and Mental Gymnastics

1. $\operatorname{cosec} = 1/\sin$, $\cot = c/s$
2. *rearrange and simplify.*

$$\text{LHS} = \operatorname{cosec} A(\operatorname{cosec} A - \sin A) + \frac{\sin A - \cos A}{\sin A} + \cot A$$

$$= \frac{1}{\sin A}\left(\frac{1}{\sin A} - \sin A\right) + \frac{\sin A - \cos A}{\sin A} + \frac{\cos A}{\sin A}$$

$$= \frac{1 - \sin^2 A}{\sin A \cdot \sin A} + \frac{\sin A - \cos A}{\sin A} + \frac{\cos A}{\sin A}$$

$$= \frac{1 - \sin^2 A + \sin^2 A - \cos A \sin A + \cos A \sin A}{\sin^2 A}$$

$$= \frac{1}{\sin^2 A}$$

$$= \operatorname{cosec}^2 A$$

$$\equiv \text{RHS}.$$

$$\sec^4 A - \tan^4 A \equiv \frac{1 + \sin^2 A}{\cos^2 A}$$

Eyeballing and Mental Gymnastics

1. $\sec = 1/\cos$, $t = s/c$
2. $\sec^4 - \tan^4$ *suggests* $a^4 - b^4 = (a^2 - b^2)(a^2 + b^2)$
3. s^2, c^2 *suggest* $s^2 + c^2 \equiv 1$
4. *rearrange and simplify.*

$$\text{LHS} = \sec^4 A - \tan^4 A$$

$$= \frac{1}{\cos^4 A} - \frac{\sin^4 A}{\cos^4 A}$$

$$= \frac{1 - \sin^4 A}{\cos^4 A}$$

$$= \frac{(1 - \sin^2 A)(1 + \sin^2 A)}{\cos^4 A}$$

$$= \frac{(\cos^2 A)(1 + \sin^2 A)}{\cos^4 A}$$

$$= \frac{1 + \sin^2 A}{\cos^2 A}$$

$$\equiv \text{RHS.}$$

$$\boxed{\frac{1 - \sin A}{\sec A} \equiv \frac{\cos^3 A}{1 + \sin A}}$$

Eyeballing and Mental Gymnastics

1. $\sec = 1/\cos$
2. $\cos^3$ suggests $c \cdot c^2$, $s^2 + c^2 \equiv 1$
3. *rearrange and simplify.*

$$\text{LHS} = \frac{1 - \sin A}{\sec A}$$

$$= (1 - \sin A) \cos A$$

$$= (1 - \sin A) \cos A \left(\frac{1 + \sin A}{1 + \sin A} \right)$$

$$= \frac{(1 - \sin^2 A) \cos A}{(1 + \sin A)}$$

$$= \frac{(\cos^2 A) \cos A}{1 + \sin A}$$

$$= \frac{\cos^3 A}{1 + \sin A}$$

$$\equiv \text{RHS.}$$

$$\boxed{\frac{(1-\cos A)}{(1+\cos A)} \equiv (\operatorname{cosec} A - \cot A)^2}$$

Eyeballing and Mental Gymnastics

1. $\operatorname{cosec} = 1/\sin$, $\cot = c/s$
2. *square on RHS suggests* $s^2 + c^2 \equiv 1$
3. *rearrange and simplify.*

$$\text{LHS} = \frac{1-\cos A}{1+\cos A}$$

$$= \frac{1-\cos A}{1+\cos A} \cdot \left(\frac{1-\cos A}{1-\cos A}\right)$$

$$= \frac{(1-\cos A)^2}{1-\cos^2 A}$$

$$= \frac{(1-\cos A)^2}{\sin^2 A}$$

$$= \left(\frac{1}{\sin} - \frac{\cos A}{\sin A}\right)^2$$

$$= (\operatorname{cosec} A - \cot A)^2$$

$$\equiv \text{RHS}.$$

$$\frac{\tan A - \cot A}{\tan A + \cot A} + 1 \equiv 2\sin^2 A$$

Eyeballing and Mental Gymnastics

1. $t = s/c$, $\cot = c/s$
2. *common denominators*
3. $s^2 + c^2 \equiv 1$
4. *rearrange and simplify.*

$$\text{LHS} = \frac{\tan A - \cot A}{\tan A + \cot A} + 1$$

$$= \frac{\dfrac{\sin A}{\cos A} - \dfrac{\cos A}{\sin A}}{\dfrac{\sin A}{\cos A} + \dfrac{\cos A}{\sin A}} + 1$$

$$= \frac{\dfrac{\sin^2 A - \cos^2 A}{\cos A \sin A}}{\dfrac{\sin^2 A + \cos^2 A}{\cos A \sin A}} + 1$$

$$= \sin^2 A - \cos^2 A + (\sin^2 A + \cos^2 A)$$

$$= 2\sin^2 A$$

$$\equiv \text{RHS}.$$

$$\boxed{\frac{1 - \cot^2 A}{1 + \cot^2 A} + 2\cos^2 A \equiv 1}$$

Eyeballing and Mental Gymnastics

1. $\cot = c/s$
2. $s^2 + c^2 \equiv 1$
3. *rearrange and simplify.*

$$\text{LHS} = \frac{1 - \cot^2 A}{1 + \cot^2 A} + 2\cos^2 A$$

$$= \frac{1 - \dfrac{\cos^2 A}{\sin^2 A}}{1 + \dfrac{\cos^2 A}{\sin^2 A}} + 2\cos^2 A$$

$$= \frac{\dfrac{\sin^2 A - \cos^2 A}{\sin^2 A}}{\dfrac{\sin^2 A + \cos^2 A}{\sin^2 A}} + 2\cos^2 A$$

$$= \frac{\sin^2 A - \cos^2 A}{1} + 2\cos^2 A$$

$$= \sin^2 A + \cos^2 A$$

$$= 1$$

$$\equiv \text{RHS.}$$

$$\boxed{\frac{\sin^2 A - \tan A}{\cos^2 A - \cot A} \equiv \tan^2 A}$$

Eyeballing and Mental Gymnastics

1. $t = s/c,\ \cot = c/s$
2. $s^2 + c^2 \equiv 1$
3. *rearrange and simplify.*

$$\text{LHS} = \frac{\sin^2 A - \tan A}{\cos^2 A - \cot A}$$

$$= \frac{\sin^2 A - \dfrac{\sin A}{\cos A}}{\cos^2 A - \dfrac{\cos A}{\sin A}}$$

$$= \frac{\sin^2 A \cos A - \sin A}{\cos A} \cdot \frac{\sin A}{\cos^2 A \sin A - \cos A}$$

$$= \frac{\sin A(\sin A \cos A - 1)}{\cos A} \cdot \frac{\sin A}{\cos A(\sin A \cos A - 1)}$$

$$= \frac{\sin^2 A}{\cos^2 A}$$

$$= \tan^2 A$$

$$\equiv \text{RHS.}$$

$$\boxed{\frac{\sec A}{1 - \sin A} \equiv \frac{1 + \sin A}{\cos^3 A}}$$

Eyeballing and Mental Gymnastics

1. $\sec = 1/\cos$
2. $s^2 + c^2 \equiv 1$
3. *rearrange and simplify.*

$$\text{LHS} = \frac{\sec A}{1 - \sin A}$$

$$= \frac{1}{\cos A} \cdot \frac{1}{(1 - \sin A)}$$

$$= \frac{1}{\cos A(1 - \sin A)} \cdot \left(\frac{1 + \sin A}{1 + \sin A}\right)$$

$$= \frac{1 + \sin A}{\cos A(1 - \sin^2 A)}$$

$$= \frac{1 + \sin A}{\cos A(\cos^2 A)}$$

$$= \frac{1 + \sin A}{\cos^3 A}$$

$$\equiv \text{RHS.}$$

$$\frac{\sec A}{\operatorname{cosec}^2 A} - \frac{\operatorname{cosec} A}{\sec^2 A} \equiv (1 + \cot A + \tan A)(\sin A - \cos A)$$

Eyeballing and Mental Gymnastics

1. cosec^2, $\sec^2$ *suggest* $s^2 + c^2 \equiv 1$
2. $\operatorname{cosec} = 1/\sin$, $\sec = 1/\cos$, $\cot = c/s$, $t = s/c$
3. *rearrange and simplify.*

$$\text{LHS} = \frac{\sec A}{\operatorname{cosec}^2 A} - \frac{\operatorname{cosec} A}{\sec^2 A}$$

$$= \frac{1}{\cos A} \cdot \left(\frac{\sin^2 A}{1}\right) - \frac{1}{\sin A}\left(\frac{\cos^2 A}{1}\right)$$

$$= \frac{\sin^2 A}{\cos A} - \frac{\cos^2 A}{\sin A}$$

$$= \frac{\sin^3 A - \cos^3 A^*}{\sin A \cos A}$$

$$= \frac{(\sin A - \cos A)(\sin^2 A + \sin A \cos A + \cos^2 A)}{\sin A \cos A}$$

$$= (\sin A - \cos A)(\tan A + 1 + \cot A)$$

$$= (1 + \cot A + \tan A)(\sin A - \cos A)$$

$$\equiv \text{RHS}.$$

---

*$(a^3 - b^3) = (a - b)(a^2 + ab + b^2)$.

$$\frac{1 + \sec A}{\sec A} \equiv \frac{\sin^2 A}{1 - \cos A}$$

Eyeballing and Mental Gymnastics

1. $\sec = 1/\cos$
2. s^2 suggests $s^2 + c^2 = 1$
3. *rearrange and simplify.*

$$\text{LHS} = \frac{1 + \sec A}{\sec A}$$

$$= \frac{1 + \dfrac{1}{\cos A}}{\dfrac{1}{\cos A}}$$

$$= \frac{1 + \cos A}{\cos A} \cdot \frac{\cos A}{1}$$

$$= 1 + \cos A$$

$$= (1 + \cos A)\left(\frac{1 - \cos A}{1 - \cos A}\right)$$

$$= \frac{1 - \cos^2 A}{1 - \cos A}$$

$$= \frac{\sin^2 A}{1 - \cos A}$$

$$\equiv \text{RHS.}$$

$$\boxed{\frac{\tan A - \cot A}{\tan A + \cot A} \equiv 1 - 2\cos^2 A}$$

Eyeballing and Mental Gymnastics

1. $t = s/c$, $\cot = c/s$
2. *common denominators*
3. $s^2 + c^2 \equiv 1$
4. *rearrange and simplify.*

$$\text{LHS} = \frac{\tan A - \cot A}{\tan A + \cot A}$$

$$= \frac{\dfrac{\sin A}{\cos A} - \dfrac{\cos A}{\sin A}}{\dfrac{\sin A}{\cos A} + \dfrac{\cos A}{\sin A}}$$

$$= \frac{\dfrac{\sin^2 A - \cos^2 A}{\cos A \sin A}}{\dfrac{\sin^2 A + \cos^2 A}{\cos A \sin A}}$$

$$= \frac{\sin^2 A - \cos^2 A}{\sin^2 A + \cos^2 A}$$

$$= \sin^2 A - \cos^2 A$$

$$= (1 - \cos^2 A) - \cos^2 A$$

$$= 1 - 2\cos^2 A$$

$$\equiv \text{RHS}.$$

$$\frac{(2\cos^2 A - 1)^2}{\cos^4 A - \sin^4 A} \equiv 1 - 2\sin^2 A$$

Eyeballing and Mental Gymnastics

1. $c^4 - s^4$ suggests $(c^2 - s^2)(c^2 + s^2)$
2. $s^2 + c^2 \equiv 1$
3. *rearrange and simplify.*

$$\text{LHS} = \frac{(2\cos^2 A - 1)^2}{\cos^4 A - \sin^4 A}$$

$$= \frac{(\cos^2 A - \sin^2 A)^2}{(\cos^2 A - \sin^2 A)(\cos^2 A + \sin^2 A)}$$

$$= \cos^2 A - \sin^2 A$$

$$= (1 - \sin^2 A) - \sin^2 A$$

$$= 1 - 2\sin^2 A$$

$$\equiv \text{RHS.}$$

$$\frac{1 - \tan^2 A}{1 + \tan^2 A} + 1 \equiv 2\cos^2 A$$

Eyeballing and Mental Gymnastics

1. $t = s/c$
2. $s^2 + c^2 \equiv 1$
3. *rearrange and simplify.*

$$\text{LHS} = \frac{1 - \tan^2 A}{1 + \tan^2 A} + 1$$

$$= \frac{1 - \dfrac{\sin^2 A}{\cos^2 A}}{1 + \dfrac{\sin^2 A}{\cos^2 A}} + 1$$

$$= \frac{\dfrac{\cos^2 A - \sin^2 A}{\cos^2 A}}{\dfrac{\cos^2 A + \sin^2 A}{\cos^2 A}} + 1$$

$$= \frac{\cos^2 A - \sin^2 A}{\cos^2 A + \sin^2 1} + 1$$

$$= \cos^2 A - \sin^2 A + 1$$

$$= \cos^2 A + \cos^2 A$$

$$= 2\cos^2 A$$

$$\equiv \text{RHS.}$$

$$\boxed{\frac{1-\sin A}{1+\sin A} \equiv (\sec A - \tan A)^2}$$

Eyeballing and Mental Gymnastics

1. $\sec = 1/\cos, t = s/c$
2. *the RHS appears to be more complex and hence should be expanded first*
3. $s^2 + c^2 \equiv 1$
4. *rearrange and simplify.*

$$\text{RHS} = (\sec A - \tan A)^2$$

$$= \left(\frac{1}{\cos A} - \frac{\sin A}{\cos A}\right)^2$$

$$= \left(\frac{1-\sin A}{\cos A}\right)^2$$

$$= \frac{(1-\sin A)^2}{\cos^2 A}$$

$$= \frac{(1-\sin A)^2}{1-\sin^2 A}$$

$$= \frac{(1-\sin A)^2}{(1-\sin A)(1+\sin A)}$$

$$= \frac{1-\sin A}{1+\sin A}$$

$$\equiv \text{LHS.}$$

$$\boxed{\frac{(\sec A - \tan A)^2 + 1}{\operatorname{cosec} A(\sec A - \tan A)} \equiv 2\tan A}$$

Eyeballing and Mental Gymnastics

1. $\sec = 1/\cos$, $t = s/c$, $\operatorname{cosec} = 1/\sin$
2. $(\quad)^2$ *suggests expansion*
3. $s^2 + c^2 \equiv 1$
4. *rearrange and simplify.*

$$\text{LHS} = \frac{(\sec A - \tan A)^2 + 1}{\operatorname{cosec} A(\sec A - \tan A)}$$

$$= \frac{\sec^2 A - 2\sec A \tan A + \tan^2 A + 1}{\operatorname{cosec} A(\sec A - \tan A)}$$

$$= \frac{\sec^2 A - 2\sec A \tan A + \sec^2 A}{\operatorname{cosec} A(\sec A - \tan A)}$$

$$= \frac{2\sec^2 A - 2\sec A \tan A}{\operatorname{cosec} A(\sec A - \tan A)}$$

$$= \frac{2\sec A(\sec A - \tan A)}{\operatorname{cosec} A(\sec A - \tan A)}$$

$$= \frac{2\sec A}{\operatorname{cosec} A}$$

$$= 2\frac{1}{\cos A} \cdot \frac{\sin A}{1}$$

$$= 2\tan A$$

$$\equiv \text{RHS.}$$

Level-Two-Games
Less-Easy Proofs

$$\boxed{\frac{2}{\cot A \tan 2A} \equiv 1 - \tan^2 A}$$

Eyeballing and Mental Gymnastics

1. *2A suggests expansion of "double angle"*
2. $\cot = 1/\tan$
3. *rearrange and simplify.*

$$\text{LHS} = \frac{2}{\cot A \tan 2A}$$

$$= 2 \tan A \cdot \frac{(1 - \tan^2 A)}{2 \tan A}$$

$$= 1 - \tan^2 A$$

$$\equiv \text{RHS.}$$

$$\boxed{\cos^4 A - \sin^4 A \equiv \frac{1}{\sec 2A}}$$

Eyeballing and Mental Gymnastics

1. $(\cos^4 - \sin^4)$ *suggests* $(a^2 - b^2)(a^2 + b^2)$
2. $\sec = 1/\cos$
3. *2A suggests "double angle"*
4. *rearrange and simplify.*

$$\text{LHS} = \cos^4 A - \sin^4 A$$

$$= (\cos^2 A - \sin^2 A)(\cos^2 A + \sin^2 A)$$

$$= (\cos 2A)(1)$$

$$= \frac{1}{\sec 2A}$$

$$\equiv \text{RHS.}$$

$$\boxed{\frac{\sin(A-B)}{\sin A \cos B} \equiv 1 - \cot A \tan B}$$

Eyeballing and Mental Gymnastics

1. $\sin(A-B)$ suggests expansion
2. $\cot = c/s, t = s/c$
3. rearrange and simplify.

$$\text{LHS} = \frac{\sin(A-B)}{\sin A \cos B}$$

$$= \frac{\sin A \cos B - \cos A \sin B}{\sin A \cos B}$$

$$= 1 - \cot A \tan B$$

$$\equiv \text{RHS}.$$

$$\frac{\cos(A-B)}{\cos A \cos B} \equiv 1 + \tan A \tan B$$

Eyeballing and Mental Gymnastics

1. $\cos(A-B)$ *suggests expansion*
2. $t = s/c$
3. *rearrange and simplify.*

$$\begin{aligned}
\text{LHS} &= \frac{\cos(A-B)}{\cos A \cos B} \\
&= \frac{\cos A \cos B + \sin A \sin B}{\cos A \cos B} \\
&= 1 + \tan A \tan B \\
&\equiv \text{RHS.}
\end{aligned}$$

$$\boxed{\frac{\cos(A+B)}{\sin A \cos B} \equiv \cot A - \tan B}$$

Eyeballing and Mental Gymnastics

1. $\cos(A+B)$ *suggests expansion*
2. $\cot = c/s, t = s/c$
3. *rearrange and simplify.*

$$\begin{aligned}
\text{LHS} &= \frac{\cos(A+B)}{\sin A \cos B} \\
&= \frac{\cos A \cos B - \sin A \sin B}{\sin A \cos B} \\
&= \cot A - \tan B \\
&\equiv \text{RHS.}
\end{aligned}$$

$$\boxed{\frac{\cos(A - B)}{\sin A \cos B} \equiv \cot A + \tan B}$$

Eyeballing and Mental Gymnastics

1. $\cos(A - B)$ *suggests expansion*
2. $\cot = c/s, t = s/c$
3. *rearrange and simplify.*

$$\text{LHS} = \frac{\cos(A - B)}{\sin A \cos B}$$

$$= \frac{\cos A \cos B + \sin A \sin B}{\sin A \cos B}$$

$$= \cot A + \tan B$$

$$\equiv \text{RHS}.$$

$$\frac{\cos(A-B)}{\sin A \sin B} \equiv 1 + \cot A \cot B$$

Eyeballing and Mental Gymnastics

1. $\cos(A-B)$ *suggests expansion*
2. $\cot = c/s$
3. *rearrange and simplify.*

$$\text{LHS} = \frac{\cos(A-B)}{\sin A \sin B}$$

$$= \frac{\cos A \cos B + \sin A \sin B}{\sin A \sin B}$$

$$= \frac{\cos A \cos B}{\sin A \sin B} + 1$$

$$= 1 + \cot A \cot B$$

$$\equiv \text{RHS.}$$

$$\frac{\sin(A+B)}{\cos A \cos B} \equiv \tan A + \tan B$$

Eyeballing and Mental Gymnastics

1. $\sin(A+B)$ *suggests expansion*
2. $t = s/c$
3. *rearrange and simplify.*

$$\text{LHS} = \frac{\sin(A+B)}{\cos A \cos B}$$

$$= \frac{\sin A \cos B + \cos A \sin B}{\cos A \cos B}$$

$$= \frac{\sin A}{\cos A} + \frac{\sin B}{\cos B}$$

$$= \tan A + \tan B$$

$$\equiv \text{RHS.}$$

$$\boxed{\cos^4 A - \sin^4 A \equiv \cos 2A}$$

Eyeballing and Mental Gymnastics

1. $c^4 - s^4$ suggests $(a^2 - b^2)(a^2 + b^2)$
2. $s^2 + c^2 \equiv 1$
3. *rearrange and simplify.*

$$\text{LHS} = \cos^4 A - \sin^4 A$$

$$= (\cos^2 A - \sin^2 A)(\cos^2 A + \sin^2 A)$$

$$= (\cos 2A)(1)$$

$$= \cos 2A$$

$$\equiv \text{RHS}.$$

$$(2a \sin A \cos A)^2 + a^2 (\cos^2 A - \sin^2 A)^2 \equiv a^2$$

Eyeballing and Mental Gymnastics

1. $2 \sin A \cos A$ *suggests* $\sin 2A$
2. $\cos^2 A - \sin^2 A$ *suggests* $\cos 2A$
3. $s^2 + c^2 \equiv 1$
4. *rearrange and simplify.*

$$\begin{aligned}
\text{LHS} &= (2a \sin A \cos A)^2 + a^2 (\cos^2 A - \sin^2 A)^2 \\
&= a^2 (\sin 2A)^2 + a^2 (\cos 2A)^2 \\
&= a^2 (\sin^2 2A + \cos^2 2A) \\
&= a^2 \\
&\equiv \text{RHS.}
\end{aligned}$$

$$\frac{\sin A}{1 + \cos A} \equiv \tan \frac{A}{2}$$

Eyeballing and Mental Gymnastics

1. *$A/2$ suggests use of "half angle formulas" for $\sin A$, $\cos A$*
2. *$t = s/c$*
3. *rearrange and simplify.*

$$\text{LHS} = \frac{\sin A}{1 + \cos A}$$

$$= \frac{2 \sin \dfrac{A}{2} \cos \dfrac{A}{2}}{1 + \left(2 \cos^2 \dfrac{A}{2} - 1 \right)}$$

$$= \frac{2 \sin \dfrac{A}{2} \cos \dfrac{A}{2}}{2 \cos^2 \dfrac{A}{2}}$$

$$= \tan \frac{A}{2}$$

$$\equiv \text{RHS.}$$

$$\boxed{\frac{1 - \cos A}{\sin A} \equiv \tan \frac{A}{2}}$$

Eyeballing and Mental Gymnastics

1. *A/2 suggests "half angle formulas" for* cos A, sin A
2. *rearrange and simplify.*

$$\text{LHS} = \frac{1 - \cos A}{\sin A}$$

$$= \frac{1 - \left(1 - 2\sin^2\dfrac{A}{2}\right)}{2\sin\dfrac{A}{2}\cos\dfrac{A}{2}}$$

$$= \frac{2\sin^2\dfrac{A}{2}}{2\sin\dfrac{A}{2}\cos\dfrac{A}{2}}$$

$$= \frac{\sin\dfrac{A}{2}}{\cos\dfrac{A}{2}}$$

$$= \tan\frac{A}{2}$$

$$\equiv \text{RHS.}$$

$$\left(\sin\frac{A}{2}+\cos\frac{A}{2}\right)^2 \equiv 1+\sin A$$

Eyeballing and Mental Gymnastics

1. *$A/2$ suggests "half angle formulas"*
2. *$(\ )^2$ suggests $s^2+c^2 \equiv 1$*
3. *rearrange and simplify.*

$$\begin{aligned}
\text{LHS} &= \left(\sin\frac{A}{2}+\cos\frac{A}{2}\right)^2 \\
&= \sin^2\frac{A}{2}+2\sin\frac{A}{2}\cos\frac{A}{2}+\cos^2\frac{A}{2} \\
&= \sin^2\frac{A}{2}+\cos^2\frac{A}{2}+2\sin\frac{A}{2}\cos\frac{A}{2} \\
&= 1+\sin A \\
&\equiv \text{RHS.}
\end{aligned}$$

$$\boxed{\cos(60° + A) + \sin(30° + A) \equiv \cos A}$$

Eyeballing and Mental Gymnastics

1. $(60° + A)$, $(30° + A)$ *suggest expansion of "compound angles"*
2. *numerical values for* sin, cos *of* $60°$, $30°$
3. *rearrange and simplify.*

$$\text{LHS} = \cos(60° + A) + \sin(30° + A)$$

$$= \cos 60° \cos A - \sin 60° \sin A$$

$$+ \sin 30° \cos A + \cos 30° \sin A$$

$$= \frac{1}{2} \cos A - \frac{\sqrt{3}}{2} \sin A + \frac{1}{2} \cos A + \frac{\sqrt{3}}{2} \sin A$$

$$= \cos A$$

$$\equiv \text{RHS.}$$

	sin	cos
0°	$0 = \sqrt{\dfrac{0}{4}}$	$\sqrt{\dfrac{4}{4}} = 1$
30°	$\dfrac{1}{2} = \sqrt{\dfrac{1}{4}}$	$\sqrt{\dfrac{3}{4}} = \dfrac{\sqrt{3}}{2}$
45°	$\dfrac{\sqrt{2}}{2} = \sqrt{\dfrac{2}{4}}$	$\sqrt{\dfrac{2}{4}} = \dfrac{\sqrt{2}}{2}$
60°	$\dfrac{\sqrt{3}}{2} = \sqrt{\dfrac{3}{4}}$	$\sqrt{\dfrac{1}{4}} = \dfrac{1}{2}$
90°	$1 = \sqrt{\dfrac{4}{4}}$	$\sqrt{\dfrac{0}{4}} = 0$

$$\frac{1 + \sin 2A + \cos 2A}{\sin A + \cos A} \equiv 2\cos A$$

Eyeballing and Mental Gymnastics

1. *2A suggests expansion of "double angle"*
2. *rearrange and simplify.*

$$\text{LHS} = \frac{1 + \sin 2A + \cos 2A}{\sin A + \cos A}$$

$$= \frac{1 + 2\sin A \cos A + (2\cos^2 A - 1)}{\sin A + \cos A}$$

$$\frac{2\cos A(\sin A + \cos A)}{\sin A + \cos A}$$

$$= 2\cos A$$

$$\equiv \text{RHS.}$$

$$\boxed{\frac{\sin 2A}{1 - \cos 2A} \equiv \cot A}$$

Eyeballing and Mental Gymnastics

1. *2A suggests expansion of "double angle"*
2. *$s^2 + c^2 \equiv 1$, $\cot = c/s$*
3. *rearrange and simplify.*

$$
\begin{aligned}
\text{LHS} &= \frac{\sin 2A}{1 - \cos 2A} \\[2mm]
&= \frac{2 \sin A \cos A}{1 - (1 - 2\sin^2 A)} \\[2mm]
&= \frac{2 \sin A \cos A}{2 \sin^2 A} \\[2mm]
&= \frac{\cos A}{\sin A} \\[2mm]
&= \cot A \\[2mm]
&\equiv \text{RHS}.
\end{aligned}
$$

$$\boxed{\frac{\sin A + \sin 2A}{1 + \cos A + \cos 2A} \equiv \tan A}$$

Eyeballing and Mental Gymnastics

1. *2A suggests expansion of "double angle"*
2. $t = s/c$
3. *rearrange and simplify.*

$$\text{LHS'} = \frac{\sin A + \sin 2A}{1 + \cos A + \cos 2A}$$

$$= \frac{\sin A + 2\sin A \cos A}{1 + \cos A + (2\cos^2 A - 1)}$$

$$= \frac{\sin A(1 + 2\cos A)}{\cos A + 2\cos^2 A}$$

$$= \frac{\sin A(1 + 2\cos A)}{\cos A(1 + 2\cos A)}$$

$$= \tan A$$

$$\equiv \text{RHS.}$$

$$\boxed{\frac{1}{2}(\cot A - \tan A) \equiv \cot 2A}$$

Eyeballing and Mental Gymnastics

1. $\cot = c/s, t = s/c$
2. $\cot 2A = \cos 2A / \sin 2A$
3. *rearrange and simplify.*

$$
\begin{aligned}
\text{LHS} &= \frac{1}{2}(\cot A - \tan A) \\
&= \frac{1}{2}\left(\frac{\cos A}{\sin A} - \frac{\sin A}{\cos A}\right) \\
&= \frac{1}{2}\left(\frac{\cos^2 - \sin^2 A}{\sin A \cos A}\right) \\
&= \frac{\cos 2A}{\sin 2A} \\
&= \cot 2A \\
&\equiv \text{RHS.}
\end{aligned}
$$

$$\boxed{\operatorname{cosec} A \sec A \equiv 2 \operatorname{cosec} 2A}$$

Eyeballing and Mental Gymnastics

1. $\operatorname{cosec} = 1/\sin$, $\sec = 1/\cos$
2. $\operatorname{cosec} 2A$ *suggests* $1/\sin 2A$
3. *rearrange and simplify.*

$$\begin{aligned}
\text{LHS} &= \operatorname{cosec} A \sec A \\
&= \frac{1}{\sin A} \cdot \frac{1}{\cos A} \\
&= \frac{1}{\sin A \cos A} \cdot \left(\frac{2}{2}\right) \\
&= \frac{2}{2\sin A \cos A} \\
&= \frac{2}{\sin 2A} \\
&= 2 \operatorname{cosec} 2A \\
&\equiv \text{RHS.}
\end{aligned}$$

$$\boxed{\frac{\cos(A+B)}{\cos A \sin B} \equiv \cot B - \tan A}$$

Eyeballing and Mental Gymnastics

1. $\cos(A+B)$ *suggests expansion*
2. $\cot = c/s$
3. *rearrange and simplify.*

$$\text{LHS} = \frac{\cos(A+B)}{\cos A \sin B}$$

$$= \frac{\cos A \cos B - \sin A \sin B}{\cos A \sin B}$$

$$= \cot B - \tan A$$

$$\equiv \text{RHS}.$$

$$\boxed{\frac{\sec A \sec B}{1 + \tan A \tan B} \equiv \sec(A - B)}$$

Eyeballing and Mental Gymnastics

1. $\sec(A - B)$ *suggests expansion*
2. $\sec = 1/\cos, t = s/c$
3. *begin with RHS since* $\sec(A - B) = 1/\cos(A - B)$ *which is standard "compound angle" function*
4. *rearrange and simplify.*

$$\text{RHS} = \sec(A - B)$$

$$= \frac{1}{\cos(A - B)}$$

$$= \frac{1}{\cos A \cos B + \sin A \sin B} \qquad \begin{array}{l} \text{divide both} \\ \text{numerator} \\ \text{and denominator} \\ \text{by } \cos A \cos B \end{array}$$

$$= \frac{\sec A \sec B}{1 + \tan A \tan B}$$

$$\equiv \text{LHS.}$$

$$\boxed{\dfrac{\operatorname{cosec} A \operatorname{cosec} B}{\cot A \cot B - 1} \equiv \sec(A + B)}$$

Eyeballing and Mental Gymnastics

1. $\sec(A + B)$ *suggests expansion*
2. $\operatorname{cosec} = 1/\sin$, $\sec = 1/\cos$, $\cot = c/s$
3. *begin with RHS since* $\sec(A + B) = 1/\cos(A + B)$ *which is a standard "compound angle" function.*
4. *rearrange and simplify.*

$$\text{RHS} = \sec(A + B)$$

$$= \dfrac{1}{\cos(A + B)}$$

$$= \dfrac{1}{\cos A \cos B - \sin A \sin B}$$ divide both numerator and denominator by $\sin A \sin B$

$$= \dfrac{\operatorname{cosec} A \operatorname{cosec} B}{\cot A \cot B - 1}$$

$$\equiv \text{LHS.}$$

$$\frac{\cot A \cot B + 1}{\cot B - \cot A} \equiv \cot(A - B)$$

Eyeballing and Mental Gymnastics

1. $\cot(A - B)$ *suggests expansion*
2. $\cot = 1/\tan$
3. *begin with RHS since* $\cot(A - B) = 1/\tan(A - B)$, *and* $\tan(A - B)$ *is a standard "compound angle" function*
4. *rearrange and simplify.*

$$\text{RHS} = \cot(A - B)$$

$$= \frac{1}{\tan(A - B)}$$

$$= \frac{1 + \tan A \tan B}{\tan A - \tan B}$$

$$= \frac{\cot A \cot B + 1}{\cot B - \cot A} \qquad \left| \begin{array}{l} \text{divide both} \\ \text{numerator} \\ \text{and denominator} \\ \text{by } \tan A \tan B \end{array} \right.$$

$$\equiv \text{LHS.}$$

$$\boxed{\frac{\cos(A+B)}{\cos A \cos B} \equiv 1 - \tan A \tan B}$$

Eyeballing and Mental Gymnastics

1. $\cos(A+B)$ suggests expansion
2. $t = s/c$
3. rearrange and simplify.

$$\begin{aligned} \text{LHS} &= \frac{\cos(A+B)}{\cos A \cos B} \\ &= \frac{\cos A \cos B - \sin A \sin B}{\cos A \cos B} \\ &= 1 - \frac{\sin A \sin B}{\cos A \cos B} \\ &= 1 - \tan A \tan B \\ &\equiv \text{RHS.} \end{aligned}$$

$$\boxed{\frac{\sin(A+B)}{\sin A \cos B} \equiv 1 + \cot A \tan B}$$

Eyeballing and Mental Gymnastics

1. $\sin(A+B)$ *suggests expansion*
2. $\cot = c/s, t = s/c$
3. *rearrange and simplify.*

$$\text{LHS} = \frac{\sin(A+B)}{\sin A \cos B}$$

$$= \frac{\sin A \cos B + \cos A \sin B}{\sin A \cos B}$$

$$= 1 + \frac{\cos A \sin B}{\sin A \cos B}$$

$$= 1 + \cot A \tan B$$

$$\equiv \text{RHS.}$$

$$\boxed{\frac{1 - \tan A \tan B}{1 + \tan A \tan B} \equiv \frac{\cos(A+B)}{\cos(A-B)}}$$

Eyeballing and Mental Gymnastics

1. $\cos(A+B)$, $\cos(A-B)$ suggest expansion
2. $t = s/c$
3. begin with RHS which are standard "compound angle" functions
4. rearrange and simplify.

$$\text{RHS} = \frac{\cos(A+B)}{\cos(A-B)}$$

$$= \frac{\cos A \cos B - \sin A \sin B}{\cos A \cos B + \sin A \sin B} \qquad \text{divide both numerator and denominator by } \cos A \cos B$$

$$= \frac{1 - \tan A \tan B}{1 + \tan A \tan B}$$

$$\equiv \text{LHS.}$$

$$\frac{\tan A + \tan B}{\tan A - \tan B} \equiv \frac{\sin(A+B)}{\sin(A-B)}$$

Eyeballing and Mental Gymnastics

1. $\sin(A+B)$, $\sin(A-B)$ suggest expansion
2. $t = s/c$
3. begin with RHS which are standard "compound angle" functions
4. rearrange and simplify.

$$\text{RHS} = \frac{\sin(A+B)}{\sin(A-B)}$$

$$= \frac{\sin A \cos B + \cos A \sin B}{\sin A \cos B - \cos A \sin B}$$

$$= \frac{\tan A + \tan B}{\tan A - \tan B}$$

divide both
numerator
and denominator
by $\cos A \cos B$

$$\equiv \text{LHS}.$$

$$\frac{\cos A - \sin A}{\cos A + \sin A} + \frac{\cos A + \sin A}{\cos A - \sin A} \equiv 2 \sec 2A$$

Eyeballing and Mental Gymnastics

1. $\sec = 1/\cos$
2. $2A$ suggests "double angle"
3. *common denominator*
4. *rearrange and simplify.*

$$\text{LHS} = \frac{\cos A - \sin A}{\cos A + \sin A} + \frac{\cos A + \sin A}{\cos A - \sin A}$$

$$= \frac{(\cos A - \sin A)^2 + (\cos A + \sin A)^2}{(\cos A + \sin A)(\cos A - \sin A)}$$

$$= \frac{(\cos^2 A - 2 \sin A \cos A + \sin^2 A) + (\cos^2 A + 2 \sin A \cos A + \sin^2 A)}{\cos^2 A - \sin^2 A}$$

$$= \frac{(1) + (1)}{\cos 2A}$$

$$= 2 \sec 2A$$

$$\equiv \text{RHS}.$$

$$\boxed{\frac{\cot A - \tan A}{\cot A + \tan A} \equiv \cos 2A}$$

Eyeballing and Mental Gymnastics

1. $\cot = c/s, t = s/c$
2. *common denominators*
3. *rearrange and simplify.*

$$\text{LHS} = \frac{\cot A - \tan A}{\cot A + \tan A}$$

$$= \frac{\dfrac{\cos A}{\sin A} - \dfrac{\sin A}{\cos A}}{\dfrac{\cos A}{\sin A} + \dfrac{\sin A}{\cos A}}$$

$$= \frac{\cos^2 A - \sin^2 A}{\sin A \cos A} \cdot \frac{\sin A \cos A}{\cos^2 A + \sin^2 A}$$

$$= \frac{\cos^2 A - \sin^2 A}{\cos^2 A + \sin^2 A}$$

$$= \frac{\cos 2A}{1}$$

$$= \cos 2A$$

$$\equiv \text{RHS.}$$

$$\boxed{\frac{\cos A - \cos B}{\sin A + \sin B} + \frac{\sin A - \sin B}{\cos A + \cos B} \equiv 0}$$

Eyeballing and Mental Gymnastics

1. *common denominators*
2. *rearrange and simplify.*

$$\text{LHS} = \left(\frac{\cos A - \cos B}{\sin A + \sin B}\right) + \left(\frac{\sin A - \sin B}{\cos A + \cos B}\right)$$

$$= \frac{(\cos A - \cos B)(\cos A + \cos B) + (\sin A - \sin B)(\sin A + \sin B)}{(\sin A + \sin B)(\cos A + \cos B)}$$

$$= \frac{(\cos^2 A - \cos^2 B) + (\sin^2 A - \sin^2 B)}{(\sin A + \sin B)(\cos A + \cos B)}$$

$$= \frac{(\cos^2 A + \sin^2 A) - (\cos^2 B + \sin^2 B)}{(\sin A + \sin B)(\cos A + \cos B)}$$

$$= \frac{(1) - (1)}{(\sin A + \sin B)(\cos A + \cos B)}$$

$$= 0$$

$$\equiv \text{RHS.}$$

$$\boxed{\frac{\cos A + \sin A}{\cos A - \sin A} - \frac{\cos A - \sin A}{\cos A + \sin A} \equiv 2 \tan 2A}$$

Eyeballing and Mental Gymnastics

1. *common denominators*
2. *tan* $2A$ *suggest "double angle" formula*
3. *lots of* $\cos^2$, $\sin^2$ *suggest* $s^2 + c^2 \equiv 1$
4. *rearrange and simplify.*

$$\text{LHS} = \left(\frac{\cos A + \sin A}{\cos A - \sin A} \right) - \left(\frac{\cos A - \sin A}{\cos A + \sin A} \right)$$

$$= \frac{(\cos A + \sin A)(\cos A + \sin A) - (\cos A - \sin A)(\cos A - \sin A)}{(\cos A - \sin A)(\cos A + \sin A)}$$

$$= \frac{(\cos^2 A + 2 \sin A \cos A + \sin^2 A) - (\cos^2 A - 2 \sin A \cos A + \sin^2 A)}{\cos^2 A - \sin^2 A}$$

$$= \frac{4 \sin A \cos A}{\cos 2A}$$

$$= 2 \frac{\sin 2A}{\cos 2A}$$

$$= 2 \tan 2A$$

$$\equiv \text{RHS.}$$

$$(4 \sin A \cos A)(1 - 2 \sin^2 A) \equiv \sin 4A$$

Eyeballing and Mental Gymnastics

1. $4 \sin A \cos A = 2(2 \sin A \cos A) = 2 \sin 2A$
2. $(1 - 2 \sin^2 A) = \cos 2A$
3. *rearrange and simplify.*

$$\text{LHS} = 4 \sin A \cos A (1 - 2 \sin^2 A)$$

$$= 2(2 \sin A \cos A)(\cos 2A)$$

$$= 2(\sin 2A)(\cos 2A)$$

$$= \sin 4A$$

$$\equiv \text{RHS.}$$

$$\boxed{\frac{\sin 3A \cos A - \sin A \cos 3A}{\sin 2A} \equiv 1}$$

Eyeballing and Mental Gymnastics

1. $\sin A \cos B - \cos A \sin B = \sin(A - B)$
2. *rearrange and simplify.*

$$\text{LHS} = \frac{\sin 3A \cos A - \sin A \cos 3A}{\sin 2A}$$

$$= \frac{\sin(3A - A)}{\sin 2A}$$

$$= \frac{\sin 2A}{\sin 2A}$$

$$= 1$$

$$\equiv \text{RHS}.$$

$$\tan\left(A - \frac{\pi}{4}\right) \equiv \frac{\tan A - 1}{\tan A + 1}$$

Eyeballing and Mental Gymnastics

1. $t = s/c$
2. $(A - (\pi/4))$ *suggests expansion of "compound angle"*
3. *rearrange and simplify.*

$$\text{LHS} = \tan\left(A - \frac{\pi}{4}\right)$$

$$= \frac{\tan A - \tan\dfrac{\pi}{4}}{1 + \tan A \tan\dfrac{\pi}{4}}$$

$$= \frac{\tan A - 1}{1 + \tan A}$$

$$= \frac{\tan A - 1}{\tan A + 1}$$

$$\equiv \text{RHS.}$$

$$\tan\frac{\pi}{4} = 1$$

$$\boxed{\frac{\cos^3 A - \sin^3 A}{\cos A - \sin A} \equiv \frac{2 + \sin 2A}{2}}$$

Eyeballing and Mental Gymnastics

1. $\cos^3 - \sin^3$ *suggests* $(a^3 - b^3) = (a-b)(a^2 + ab + b^2)$
2. $\sin 2A$ *suggests "double angle"*
3. *rearrange and simplify.*

$$\text{LHS} = \frac{\cos^3 A - \sin^3 A}{\cos A - \sin A}$$

$$= \frac{(\cos A - \sin A)(\cos^2 A + \cos A \sin A + \sin^2 A)}{(\cos A - \sin A)}$$

$$= (1 + \cos A \sin A)$$

$$= \frac{1}{2}(2 + 2\cos A \sin A)$$

$$= \frac{2 + \sin 2A}{2}$$

$$\equiv \text{RHS.}$$

$$\boxed{\frac{\sin 3A}{\sin A} - \frac{\cos 3A}{\cos A} \equiv 2}$$

Eyeballing and Mental Gymnastics

1. *common denominator*
2. $\sin(A - B)$ *expansion*
3. *rearrange and simplify.*

$$\text{LHS} = \frac{\sin 3A}{\sin A} - \frac{\cos 3A}{\cos A}$$

$$= \frac{\sin 3A \cos A - \cos 3A \sin A}{\sin A \cos A}$$

$$= \frac{\sin(3A - A)}{\sin A \cos A}$$

$$= \frac{\sin 2A}{\frac{1}{2}(2 \sin A \cos A)}$$

$$= 2\frac{\sin 2A}{\sin 2A}$$

$$= 2$$

$$\equiv \text{RHS}.$$

$$\boxed{2 \cot A \cot 2A \equiv \cot^2 A - 1}$$

Eyeballing and Mental Gymnastics

1. $\cot = c/s$
2. $\cos 2A$, $\sin 2A$ *suggest expansion*
3. *rearrange and simplify.*

$$
\begin{aligned}
\text{LHS} &= 2 \cot A \cot 2A \\
&= 2 \frac{\cos A}{\sin A} \cdot \frac{\cos 2A}{\sin 2A} \\
&= 2 \frac{\cos A}{\sin A} \cdot \frac{(\cos^2 A - \sin^2 A)}{2 \sin A \cos A} \\
&= \frac{1}{\sin^2 A} (\cos^2 A - \sin^2 A) \\
&= \cot^2 A - 1 \\
&\equiv \text{RHS.}
\end{aligned}
$$

$$\boxed{\frac{1}{2} \sec A \operatorname{cosec} A \equiv \operatorname{cosec} 2A}$$

Eyeballing and Mental Gymnastics

1. $\sec = 1/\cos$, $\operatorname{cosec} = 1/\sin$
2. $\operatorname{cosec} 2A = 1/\sin 2A$
3. *rearrange and simplify.*

$$\text{LHS} = \frac{1}{2} \sec A \operatorname{cosec} A$$

$$= \frac{1}{2} \frac{1}{\cos A} \cdot \frac{1}{\sin A}$$

$$= \frac{1}{2 \sin A \cos A}$$

$$= \frac{1}{\sin 2A}$$

$$= \operatorname{cosec} 2A$$

$$\equiv \text{RHS.}$$

$$\boxed{\tan A + \cot A \equiv \frac{2}{\sin 2A}}$$

Eyeballing and Mental Gymnastics

1. $t = s/c$, $\cot = c/s$
2. $\sin 2A$ suggests *"double angle"*
3. *rearrange and simplify.*

$$\text{LHS} = \tan A + \cot A$$

$$= \frac{\sin A}{\cos A} + \frac{\cos A}{\sin A}$$

$$= \frac{\sin^2 A + \cos^2 A}{\sin A \cos A}$$

$$= \frac{1}{\sin A \cos A}$$

$$= \frac{2}{2 \sin A \cos A}$$

$$= \frac{2}{\sin 2A}$$

$$\equiv \text{RHS.}$$

$$\frac{2}{1 - \cos A} \equiv \operatorname{cosec}^2 \frac{A}{2}$$

Eyeballing and Mental Gymnastics

1. *A/2 suggests expansion of* $\cos A$
2. $\operatorname{cosec} = 1/\sin$
3. *rearrange and simplify.*

$$\text{LHS} = \frac{2}{1 - \cos A}$$

$$= \frac{2}{1 - \left(1 - 2\sin^2 \dfrac{A}{2}\right)}$$

$$= \frac{2}{2\sin^2 \dfrac{A}{2}}$$

$$= \frac{1}{\sin^2 \dfrac{A}{2}}$$

$$= \operatorname{cosec}^2 \frac{A}{2}$$

$$\equiv \text{RHS}.$$

$$\boxed{\frac{2}{1 + \cos A} \equiv \sec^2 \frac{A}{2}}$$

Eyeballing and Mental Gymnastics

1. *$A/2$ suggests expansion of* $\cos A$
2. $\sec^2 = 1/\cos^2$
3. *rearrange and simplify.*

$$\text{LHS} = \frac{2}{1 + \cos A}$$

$$= \frac{2}{1 + \left(2 \cos^2 \frac{A}{2} - 1\right)}$$

$$= \frac{2}{2 \cos^2 \frac{A}{2}}$$

$$= \frac{1}{\cos^2 \frac{A}{2}}$$

$$= \sec^2 \frac{A}{2}$$

$$\equiv \text{RHS}.$$

$$\boxed{(1 + \cos A)\tan \frac{A}{2} \equiv \sin A}$$

Eyeballing and Mental Gymnastics

1. $\tan(A/2)$ *suggests expansion of* $\cos A$ *as* $\cos 2(A/2)$ *and* $\sin A$ *as* $\sin 2(A/2)$
2. *rearrange and simplify.*

$$\text{LHS} = (1 + \cos A)\tan \frac{A}{2}$$

$$= \left(1 + \cos 2\left(\frac{A}{2}\right)\right)\tan \frac{A}{2}$$

$$= \left(1 + \left(2\cos^2 \frac{A}{2} - 1\right)\right)\tan \frac{A}{2}$$

$$= 2\cos^2 \frac{A}{2} \cdot \frac{\sin \frac{A}{2}}{\cos \frac{A}{2}}$$

$$= 2\cos \frac{A}{2}\sin \frac{A}{2}$$

$$= \sin A$$

$$\equiv \text{RHS.}$$

$$\boxed{\frac{\sin^3 A + \cos^3 A}{\sin A + \cos A} \equiv 1 - \frac{\sin 2A}{2}}$$

Eyeballing and Mental Gymnastics

1. $s^3 + c^3 = (s+c)(s^2 - sc + c^2)$
2. $s^2 + c^2 \equiv 1$
3. $\sin 2A = 2 \sin A \cos A$
4. *rearrange and simplify.*

$$\begin{aligned}
\text{LHS} &= \frac{\sin^3 A + \cos^3 A}{\sin A + \cos A} \\[2mm]
&= \frac{(\sin A + \cos A)(\sin^2 A - \sin A \cos A + \cos^2 A)}{(\sin A + \cos A)} \\[2mm]
&= \sin^2 A - \sin A \cos A + \cos^2 A \\[2mm]
&= 1 - \sin A \cos A \\[2mm]
&= 1 - \frac{1}{2}(2 \sin A \cos A) \\[2mm]
&= 1 - \frac{\sin 2A}{2} \\[2mm]
&\equiv \text{RHS.}
\end{aligned}$$

$$\boxed{\frac{\sec^2 A}{2 - \sec^2 A} \equiv \sec 2A}$$

Eyeballing and Mental Gymnastics

1. $\sec = 1/\cos$, $\sec 2A = 1/\cos 2A$
2. $\sec^2$ *suggests* $s^2 + c^2 \equiv 1$
3. *rearrange and simplify.*

$$\text{LHS} = \frac{\sec^2 A}{2 - \sec^2 A}$$

$$= \frac{1}{\cos^2 A} \cdot \frac{1}{2 - \dfrac{1}{\cos^2 A}}$$

$$= \frac{1}{\cos^2 A} \cdot \frac{\cos^2 A}{(2\cos^2 A - 1)}$$

$$= \frac{1}{2\cos^2 A - 1}$$

$$= \frac{1}{\cos 2A}$$

$$= \sec 2A$$

$$\equiv \text{RHS.}$$

$$\boxed{\operatorname{cosec} A - \cot A \equiv \tan \frac{A}{2}}$$

Eyeballing and Mental Gymnastics

1. $\operatorname{cosec} = 1/\sin,\ \cot = c/s,\ t = s/c$
2. *$A/2$ suggests expansion of* $\sin A$, $\cos A$
3. *rearrange and simplify.*

$$\text{LHS} = \operatorname{cosec} A - \cot A$$

$$= \frac{1}{\sin A} - \frac{\cos A}{\sin A}$$

$$= \frac{1 - \cos A}{\sin A}$$

$$= \frac{1 - \left(1 - 2\sin^2\frac{A}{2}\right)}{2\sin\frac{A}{2}\cos\frac{A}{2}}$$

$$= \frac{2\sin^2\frac{A}{2}}{2\sin\frac{A}{2}\cos\frac{A}{2}}$$

$$= \frac{\sin\frac{A}{2}}{\cos\frac{A}{2}}$$

$$= \tan\frac{A}{2}$$

$$\equiv \text{RHS.}$$

$$\boxed{\sin A \tan \frac{A}{2} \equiv 1 - \cos A}$$

Eyeballing and Mental Gymnastics

1. $\tan(A/2)$ *suggests expansion of* $\sin A$ *as* $\sin 2(A/2)$, *and* $\cos A$ *as* $\cos 2(A/2)$
2. $t = s/c$
3. *rearrange and simplify.*

$$\text{LHS} = \sin A \tan \frac{A}{2}$$

$$= \sin 2 \left(\frac{A}{2} \right) \tan \frac{A}{2}$$

$$= 2 \sin \frac{A}{2} \cdot \cos \frac{A}{2} \cdot \frac{\sin \frac{A}{2}}{\cos \frac{A}{2}}$$

$$= 2 \sin^2 \frac{A}{2}$$

$$= \left(2 \sin^2 \frac{A}{2} - 1 \right) + 1$$

$$= 1 - \cos A$$

$$\equiv \text{RHS.}$$

$$\boxed{\sin 2A \tan A \equiv 1 - \cos 2A}$$

Eyeballing and Mental Gymnastics

1. *Expand "double angle"*
2. $t = s/c$
3. *rearrange and simplify.*

$$
\begin{aligned}
\text{LHS} &= \sin 2A \tan A \\
&= 2 \sin A \cos A \cdot \frac{\sin A}{\cos A} \\
&= 2 \sin^2 A \\
&= (2 \sin^2 A - 1) + 1 \\
&= 1 - \cos 2A \\
&\equiv \text{RHS.}
\end{aligned}
$$

$$\boxed{\operatorname{cosec} 2A - \cot 2A \equiv \tan A}$$

Eyeballing and Mental Gymnastics

1. *2A suggests expansion of "double angle"*
2. $\operatorname{cosec} = 1/\sin, \cot = c/s, t = s/c$
3. *rearrange and simplify.*

$$\begin{aligned}
\text{LHS} &= \operatorname{cosec} 2A - \cot 2A \\
&= \frac{1}{\sin 2A} - \frac{\cos 2A}{\sin 2A} \\
&= \frac{1 - \cos 2A}{\sin 2A} \\
&= \frac{1 - (1 - 2\sin^2 A)}{2\sin A \cos A} \\
&= \frac{\sin A}{\cos A} \\
&= \tan A \\
&\equiv \text{RHS.}
\end{aligned}$$

$$\tan\left(45° + \frac{A}{2}\right) \equiv \tan A + \sec A$$

Eyeballing and Mental Gymnastics

1. $(45° + A/2)$ suggests expansion of "compound angle"
2. $A/2$ suggests "half angle" expansion
3. both sides are complex; hence explore simplification on both sides to achieve identity.

$$\text{LHS} = \tan\left(45° + \frac{A}{2}\right)$$

$$= \frac{\tan 45° + \tan\dfrac{A}{2}}{1 - \tan 45° \tan\dfrac{A}{2}}$$

$$= \frac{1 + \tan\dfrac{A}{2}}{1 - \tan\dfrac{A}{2}}$$

$$\text{RHS} = \tan A + \sec A$$

$$= \frac{\sin A}{\cos A} + \frac{1}{\cos A}$$

$$= \frac{\sin A + 1}{\cos A}$$

$$= \frac{2\sin\dfrac{A}{2}\cos\dfrac{A}{2} + \left(\cos^2\dfrac{A}{2} + \sin^2\dfrac{A}{2}\right)}{\cos^2\dfrac{A}{2} - \sin^2\dfrac{A}{2}}$$

$$= \frac{\left(\sin\dfrac{A}{2} + \cos\dfrac{A}{2}\right)^2}{\left(\cos\dfrac{A}{2} + \sin\dfrac{A}{2}\right)\left(\cos\dfrac{A}{2} - \sin\dfrac{A}{2}\right)} = \frac{\left(\sin\dfrac{A}{2} + \cos\dfrac{A}{2}\right)}{\left(\cos\dfrac{A}{2} - \sin\dfrac{A}{2}\right)}$$

$$= \frac{\tan\dfrac{A}{2} + 1}{1 - \tan\dfrac{A}{2}} \qquad \begin{array}{l}\text{divide numerator} \\ \text{and denominator} \\ \text{by } \cos\dfrac{A}{2}\end{array}$$

$$\equiv \text{simplified form of LHS.}$$

A more elegant proof comes from the use of the application of Pythagoras Theorem with respect to the "half angle formula" of $\tan A$.

$$\tan A = \frac{2\tan\dfrac{A}{2}}{1 - \tan^2 \dfrac{A}{2}}.$$

Let $t = \tan(A/2)$.
Then from Pythagoras Theorem

$$\text{the hypotenuse} = \sqrt{(2t)^2 + (1 - t^2)^2}$$

$$= \sqrt{1 + 2(2t) + t^4}$$

$$= 1 + t^2$$

$$\therefore \cos A = \frac{1 - t^2}{1 + t^2}$$

$$\therefore \text{RHS} = \tan A + \sec A$$

$$= \tan A + \frac{1}{\cos A}$$

$$= \frac{2t}{1 - t^2} + \frac{1 + t^2}{1 - t^2} \qquad \text{where } t = \tan\frac{A}{2}$$

$$= \frac{1 + 2t + t^2}{(1 - t^2)}$$

$$= \frac{(1 + t)^2}{(1 + t)(1 - t)}$$

$$= \frac{1 + t}{1 - t}$$

$$= \frac{1 + \tan\dfrac{A}{2}}{1 - \tan\dfrac{A}{2}}$$

$$\equiv \text{simplified form of LHS.}$$

$$\frac{\tan A}{\tan 2A - \tan A} \equiv \cos 2A$$

Eyeballing and Mental Gymnastics

1. $t = s/c$
2. $2A$ suggests expansion of "compound angle"
3. *rearrange and simplify.*

$$\text{LHS} = \frac{\tan A}{\tan 2A - \tan A}$$

$$= \frac{\sin A}{\cos A} \cdot \frac{1}{\left(\dfrac{\sin 2A}{\cos 2A} - \dfrac{\sin A}{\cos A}\right)}$$

$$= \frac{\sin A}{\cos A} \cdot \frac{\cos A \cos 2A}{(\sin 2A \cos A - \sin A \cos 2A)}$$

$$= \frac{\sin A \cos 2A}{(\sin 2A \cos A - \cos 2A \sin A)}$$

$$= \frac{\sin A \cos 2A}{\sin(2A - A)}$$

$$= \cos 2A$$

$$\equiv \text{RHS.}$$

$$\boxed{\sec 2A - \tan 2A \equiv \tan(45° - A)}$$

Eyeballing and Mental Gymnastics

1. $2A$, $(45° - A)$ *suggest expansion of "compound angles"*
2. *since both LHS and RHS have complex functions, explore simplification of both side to achieve identity.*
3. $\sec = 1/\cos$, $t = s/c$
4. *rearrange and simplify.*

$$\text{LHS} = \sec 2A - \tan 2A$$

$$= \frac{1}{\cos 2A} - \frac{\sin 2A}{\cos 2A}$$

$$= \frac{1 - \sin 2A}{\cos 2A}$$

$$= \frac{(\cos^2 A + \sin^2 A - 2\sin A\cos A)}{\cos^2 A - \sin^2 A}$$

$$= \frac{(\cos A - \sin A)^2}{(\cos A + \sin A)(\cos A - \sin A)}$$

$$= \frac{(\cos A - \sin A)}{(\cos A + \sin A)}$$

$$= \frac{1 - \tan A}{1 + \tan A}$$

dividing all terms by $\cos A$ to prepare for comparison with RHS.

$$\text{RHS} = \tan(45° - A)$$

$$= \frac{\tan 45° - \tan A}{1 + \tan 45° \tan A}$$

$$= \frac{1 - \tan A}{1 + \tan A}$$

$$\equiv \text{LHS}.$$

$$2\sin^2\frac{A}{6} - \sin^2\frac{A}{7} \equiv \cos^2\frac{A}{7} - \cos\frac{A}{3}$$

Eyeballing and Mental Gymnastics

1. $\sin^2(A/7)$, $\cos^2(A/7)$ suggest $s^2 + c^2 \equiv 1$
2. $A/3$ suggests "double angle" formula to give $A/6$
3. *rearrange and simplify.*

$$\text{LHS} = 2\sin^2\frac{A}{6} - \sin^2\frac{A}{7}$$

$$= 2\sin^2\frac{A}{6} - \left(1 - \cos^2\frac{A}{7}\right)$$

$$= \left(2\sin^2\frac{A}{6} - 1\right) + \cos^2\frac{A}{7}$$

$$= -\cos 2\left(\frac{A}{6}\right) + \cos^2\frac{A}{7}$$

$$= \cos^2\frac{A}{7} - \cos\frac{A}{3}$$

$$\equiv \text{RHS.}$$

$$\boxed{\frac{\sin 2A}{1 + \cos 2A} \equiv \tan A}$$

Eyeballing and Mental Gymnastics

1. $2A$ suggests expansion of "double angle"
2. rearrange and simplify.

$$\text{LHS} = \frac{\sin 2A}{1 + \cos 2A}$$

$$= \frac{2 \sin A \cos A}{1 + (2 \cos^2 A - 1)}$$

$$= \frac{2 \sin A \cos A}{2 \cos^2 A}$$

$$= \frac{\sin A}{\cos A}$$

$$= \tan A$$

$$\equiv \text{RHS}.$$

$$\boxed{\tan A + \cot A \equiv 2\operatorname{cosec} 2A}$$

Eyeballing and Mental Gymnastics

1. $t = s/c$, $\cot = c/s$, $\operatorname{cosec} = 1/\sin$
2. $2A$ suggests *"double angle"*
3. *rearrange and simplify.*

$$\text{LHS} = \tan A + \cot A$$

$$= \frac{\sin A}{\cos A} + \frac{\cos A}{\sin A}$$

$$= \frac{\sin^2 A + \cos^2 A}{\cos A \sin A}$$

$$= \frac{1}{\cos A \sin A}$$

$$= \frac{2}{2} \cdot \frac{1}{\cos A \sin A}$$

$$= \frac{2}{\sin 2A}$$

$$= 2\operatorname{cosec} 2A$$

$$\equiv \text{RHS.}$$

$$\boxed{2\cos^2(45° - A) \equiv 1 + \sin 2A}$$

Eyeballing and Mental Gymnastics

1. *c^2 suggests $s^2 + c^2 \equiv 1$*
2. *$(45° - A)$, $2A$ suggest expansion of "compound angles"*
3. *rearrange and simplify.*

$$\text{LHS} = 2\cos^2(45° - A)$$

$$= 2(\cos 45° \cos A + \sin 45° \sin A)^2$$

$$= 2\left(\frac{\sqrt{2}}{2}\cos A + \frac{\sqrt{2}}{2}\sin A\right)^2$$

$$= 2\left(\frac{\sqrt{2}}{2}\right)^2 (\cos A + \sin A)^2$$

$$= \cos^2 A + 2\sin A \cos A + \sin^2 A$$

$$= 1 + \sin 2A$$

$$\equiv \text{RHS.}$$

	sin	cos
45°	$\sqrt{\dfrac{2}{4}}$	$\sqrt{\dfrac{2}{4}}$

$$\boxed{\cos(A+B)\cos(A-B) \equiv \cos^2 B - \sin^2 A}$$

Eyeballing and Mental Gymnastics

1. *Expand "compound angles"*
2. c^2, s^2 *suggest* $s^2 + c^2 \equiv 1$
3. *rearrange and simplify.*

$$\text{LHS} = \cos(A+B)\cos(A-B)$$

$$= (\cos A \cos B - \sin A \sin B)(\cos A \cos B + \sin A \sin B)$$

$$= \cos^2 A \cos^2 B - \sin^2 A \sin^2 B$$

$$= \cos^2 B(1 - \sin^2 A) - \sin^2 A \sin^2 B$$

$$= \cos^2 B - \cos^2 B \sin^2 A - \sin^2 A \sin^2 B$$

$$= \cos^2 B - \sin^2 A(\cos^2 B + \sin^2 B)$$

$$= \cos^2 B - \sin^2 A$$

$$\equiv \text{RHS}.$$

$$\boxed{\cos(A+B)\cos(A-B) \equiv \cos^2 A - \sin^2 B}$$

Eyeballing and Mental Gymnastics

1. *Expand "compound angles"*
2. *c^2, s^2 suggest $s^2 + c^2 \equiv 1$*
3. *rearrange and simplify.*

$$\text{LHS} = \cos(A+B)\cos(A-B)$$

$$= (\cos A \cos B - \sin A \sin B)(\cos A \cos B + \sin A \sin B)$$

$$= \cos^2 A \cos^2 B - \sin^2 A \sin^2 B$$

$$= \cos^2 A(1 - \sin^2 B) - \sin^2 A \sin^2 B$$

$$= \cos^2 A - \cos^2 A \sin^2 B - \sin^2 A \sin^2 B$$

$$= \cos^2 A - \sin^2 B(\cos^2 A + \sin^2 A)$$

$$= \cos^2 A - \sin^2 B$$

$$\equiv \text{RHS.}$$

$$\sin(A+B)\sin(A-B) \equiv \sin^2 A - \sin^2 B$$

Eyeballing and Mental Gymnastics

1. *Expand "compound angles"*
2. *s^2 suggests $s^2 + c^2 \equiv 1$*
3. *rearrange and simplify.*

$$\text{LHS} = \sin(A+B)\sin(A-B)$$

$$= (\sin A \cos B + \cos A \sin B)(\sin A \cos B - \cos A \sin B)$$

$$= \sin^2 A \cos^2 B - \cos^2 A \sin^2 B$$

$$= \sin^2 A (1 - \sin^2 B) - \cos^2 A \sin^2 B$$

$$= \sin^2 A - \sin^2 A \sin^2 B - \cos^2 A \sin^2 B$$

$$= \sin^2 A - \sin^2 B(\sin^2 A + \cos^2 A)$$

$$= \sin^2 A - \sin^2 B$$

$$\equiv \text{RHS.}$$

$$\boxed{\sin(A+B)\sin(A-B) \equiv \cos^2 B - \cos^2 A}$$

Eyeballing and Mental Gymnastics

1. *Expand "compound angles"*
2. c^2 *suggests* $s^2 + c^2 \equiv 1$
3. *rearrange and simplify.*

$$\text{LHS} = \sin(A+B)\sin(A-B)$$

$$= (\sin A \cos B + \cos A \sin B)(\sin A \cos B - \cos A \sin B)$$

$$= \sin^2 A \cos^2 B - \cos^2 A \sin^2 B$$

$$= \cos^2 B(1 - \cos^2 A) - \cos^2 A \sin^2 B$$

$$= \cos^2 B - \cos^2 A \cos^2 B - \cos^2 A \sin^2 B$$

$$= \cos^2 B - \cos^2 A(\cos^2 B + \sin^2 B)$$

$$= \cos^2 B - \cos^2 A$$

$$\equiv \text{RHS.}$$

$$\boxed{\frac{\sec A + 1}{\sec A - 1} \equiv \cot^2 \frac{A}{2}}$$

Eyeballing and Mental Gymnastics

1. $\sec = 1/\cos$, $\cot = c/s$
2. $A/2$ *suggests expansion of* $\cos A$
3. $\cot^2$ *suggests* $s^2 + c^2 \equiv 1$
4. *rearrange and simplify.*

$$\text{LHS} = \frac{\sec A + 1}{\sec A - 1}$$

$$= \frac{\left(\dfrac{1}{\cos A} + 1\right)}{\left(\dfrac{1}{\cos A} - 1\right)}$$

$$= \left(\frac{1 + \cos A}{\cos A}\right) \cdot \left(\frac{\cos A}{1 - \cos A}\right)$$

$$= \frac{1 + \cos A}{1 - \cos A}$$

$$= \frac{1 + \left(2\cos^2 \dfrac{A}{2} - 1\right)}{1 - \left(1 - 2\sin^2 \dfrac{A}{2}\right)}$$

$$= \frac{2\cos^2 \dfrac{A}{2}}{2\sin^2 \dfrac{A}{2}}$$

$$= \cot^2 \frac{A}{2}$$

$$\equiv \text{RHS}.$$

$$\boxed{\frac{\sec^2 A}{2 - \sec^2 A} \equiv \sec 2A}$$

Eyeballing and Mental Gymnastics

1. $\sec = 1/\cos$
2. $2A$ suggests *"compound angle"*
3. *rearrange and simplify.*

$$\text{LHS} = \frac{\sec^2 A}{2 - \sec^2 A}$$

$$= \frac{1}{\cos^2 A} \cdot \frac{1}{\left(2 - \dfrac{1}{\cos^2 A}\right)}$$

$$= \frac{1}{\cos^2 A} \cdot \frac{\cos^2 A}{(2\cos^2 A - 1)}$$

$$= \frac{1}{\cos 2A}$$

$$= \sec 2A$$

$$\equiv \text{RHS.}$$

$$\frac{1}{8}(1 - \cos 4A) \equiv \sin^2 A \cos^2 A$$

Eyeballing and Mental Gymnastics

1. $\cos 4A$ *suggests expansion twice*: $\cos 4A \rightarrow \cos 2A \rightarrow \cos A$
2. s^2, c^2 *suggest* $s^2 + c^2 \equiv 1$
3. *rearrange and simplify.*

$$\text{LHS} = \frac{1}{8}(1 - \cos 4A)$$

$$= \frac{1}{8}(1 - (2\cos^2 2A - 1))$$

$$= \frac{1}{8}(2 - 2(2\cos^2 A - 1)^2)$$

$$= \frac{1}{8} \cdot 2(1 - (4\cos^4 A - 4\cos^2 A + 1))$$

$$= \frac{1}{4}(4\cos^2 A(\cos^2 A - 1))$$

$$= \cos^2 A(\sin^2 A)$$

$$= \sin^2 A \cos^2 A$$

$$\equiv \text{RHS.}$$

$$\boxed{\frac{\cot^2 A - 1}{2 \cot A} \equiv \cot 2A}$$

Eyeballing and Mental Gymnastics

1. $\cot = c/s, t = s/c$
2. $\cot 2A = \cos 2A / \sin 2A$
3. *rearrange and simplify.*

$$\begin{aligned}
\text{LHS} &= \frac{\cot^2 A - 1}{2 \cot A} \\[2mm]
&= \frac{1}{2}\left(\cot A - \frac{1}{\cot A} \right) \\[2mm]
&= \frac{1}{2}(\cot A - \tan A) \\[2mm]
&= \frac{1}{2}\left(\frac{\cos A}{\sin A} - \frac{\sin A}{\cos A} \right) \\[2mm]
&= \frac{1}{2}\left(\frac{\cos^2 A - \sin^2 A}{\sin A \cos A} \right) \\[2mm]
&= \frac{\cos 2A}{\sin 2A} \\[2mm]
&= \cot 2A \\[2mm]
&\equiv \text{RHS.}
\end{aligned}$$

$$\boxed{\frac{2 \tan A}{1 + \tan^2 A} \equiv \sin 2A}$$

Eyeballing and Mental Gymnastics

1. $t = s/c$
2. $2A$ suggests *"double angle"*
3. *rearrange and simplify.*

$$\text{LHS} = \frac{2 \tan A}{1 + \tan^2 A}$$

$$= \frac{2 \left(\dfrac{\sin A}{\cos A} \right)}{1 + \left(\dfrac{\sin A}{\cos A} \right)^2}$$

$$= 2 \left(\frac{\sin A}{\cos A} \right) \left(\frac{\cos^2 A}{\cos^2 A + \sin^2 A} \right)$$

$$= 2 \frac{\sin A}{\cos A} \left(\frac{\cos^2 A}{1} \right)$$

$$= 2 \sin A \cos A$$

$$= \sin 2A$$

$$\equiv \text{RHS.}$$

$$\boxed{\frac{1 - \tan^2 A}{1 + \tan^2 A} \equiv \cos 2A}$$

Eyeballing and Mental Gymnastics

1. $t = s/c$
2. $2A$ suggests *"double angle"*
3. *rearrange and simplify.*

$$\text{LHS} = \frac{1 - \tan^2 A}{1 + \tan^2 A}$$

$$= \left(1 - \left(\frac{\sin A}{\cos A}\right)^2\right) \frac{1}{\left(1 + \left(\frac{\sin A}{\cos A}\right)^2\right)}$$

$$= \left(\frac{\cos^2 A - \sin^2 A}{\cos^2 A}\right) \left(\frac{\cos^2 A}{\cos^2 A + \sin^2 A}\right)$$

$$= \frac{(\cos^2 A - \sin^2 A)}{\cos^2 A} \cdot \frac{\cos^2 A}{1}$$

$$= \cos^2 A - \sin^2 A$$

$$= \cos 2A$$

$$\equiv \text{RHS.}$$

The previous two identities are part of the series of tan $2A$, sin $2A$ and cos $2A$ in terms of tan A.

$$\tan 2A = \frac{2\tan A}{1 - \tan^2 A}$$

$$\sin 2A = \frac{2\tan A}{1 + \tan^2 A}$$

$$\cos 2A = \frac{1 - \tan^2 A}{1 + \tan^2 A}$$

The "double angle" formula for tan is well known to generations of students. But few are those who know about these special formulas for sin $2A$ and cos $2A$ in terms of tan A. You are among the very few. Test it out yourself with your friends!

An easy way to remember the three identities is to write tan A as t.

$$\text{then:} \quad \tan 2A = \frac{2t}{1 - t^2}$$

From Pythagoras Theorem, the hypotenuse is given by:

$$\sqrt{(2t)^2 + (1 - t^2)^2}$$

$$= \sqrt{4t^2 + 1 - 2t^2 + t^4}$$

$$= \sqrt{1 + 2t^2 + t^4}$$

$$= 1 + t^2$$

Then:

$$\sin 2A = \frac{O}{H} = \frac{2t}{1 + t^2}, \quad \text{and}$$

$$\cos 2A = \frac{A}{H} = \frac{1 - t^2}{1 + t^2}.$$

This sin 2*A* equation is one of the amazing equations in Trigonometry where a slight difference (from minus to plus sign in the denominator) changes the tangent identity for the double angle to the sine identity for the same double angle.

$$\frac{\cos 2A}{\sin A} + \frac{\sin 2A}{\cos A} \equiv \operatorname{cosec} A$$

Eyeballing and Mental Gymnastics

1. $2A$ suggests expansion of "double angle"
2. $\operatorname{cosec} = 1/\sin$
3. *rearrange and simplify.*

$$\text{LHS} = \frac{\cos 2A}{\sin A} + \frac{\sin 2A}{\cos A}$$

$$= \frac{\cos^2 A - \sin^2 A}{\sin A} + \frac{2 \sin A \cos A}{\cos A}$$

$$= \frac{1 - 2 \sin^2 A}{\sin A} + 2 \sin A$$

$$= \frac{1 - 2 \sin^2 A + 2 \sin^2 A}{\sin A}$$

$$= \frac{1}{\sin A}$$

$$= \operatorname{cosec} A$$

$$\equiv \text{RHS}.$$

$$\boxed{\frac{2\sin(A-B)}{\cos(A+B)-\cos(A-B)} \equiv \cot A - \cot B}$$

Eyeballing and Mental Gymnastics

1. $(A+B)$, $(A-B)$ *suggest expansion of "compound angles"*
2. $\cot = c/s$
3. *rearrange and simplify.*

$$\text{LHS} = \frac{2\sin(A-B)}{\cos(A+B)-\cos(A-B)}$$

$$= \frac{2(\sin A \cos B - \cos A \sin B)}{(\cos A \cos B - \sin A \sin B) - (\cos A \cos B + \sin A \sin B)}$$

$$= \frac{2(\sin A \cos B - \cos A \sin B)}{-2\sin A \sin B}$$

$$= \frac{(\cos A \sin B - \sin A \cos B)}{\sin A \sin B}$$

$$= \cot A - \cot B$$

$$\equiv \text{RHS}.$$

$$\frac{\cos 2A}{1 + \sin 2A} \equiv \frac{\cot A - 1}{\cot A + 1}$$

Eyeballing and Mental Gymnastics

1. $\cos 2A$, $\sin 2A$ suggest expansion
2. $\cot = c/s$
3. *rearrange and simplify.*

$$\text{LHS} = \frac{\cos 2A}{1 + \sin 2A}$$

$$= \frac{(\cos^2 A - \sin^2 A)}{1 + 2\sin A \cos A}$$

$$= \frac{\cos^2 A - \sin^2 A}{\cos^2 A + \sin^2 A + 2\sin A \cos A}$$

$$= \frac{(\cos A - \sin A)(\cos A + \sin A)}{(\cos A + \sin A)^2}$$

$$= \frac{\cos A - \sin A}{\cos A + \sin A}$$

$$= \frac{\cot A - 1}{\cot A + 1}$$

dividing both the numerator and the denominator by sin.

$$\equiv \text{RHS}.$$

$$\boxed{\frac{\cot A \cot B - 1}{\cot A + \cot B} \equiv \cot(A + B)}$$

Eyeballing and Mental Gymnastics

1. $(A + B)$ *on RHS suggests expansion*
2. $\cot = 1/\tan$
3. *begin with RHS expansion of* $\tan(A + B)$
4. *rearrange and simplify.*

$$\text{RHS} = \cot(A + B)$$

$$= \frac{1}{\tan(A + B)}$$

$$= \frac{1 - \tan A \tan B}{\tan A + \tan B}$$

$$= \frac{\cot A \cot B - 1}{\cot B + \cot A} \quad \left|\begin{array}{l}\text{divide both} \\ \text{numerator} \\ \text{and denominator} \\ \text{by } \tan A \tan B\end{array}\right.$$

$$= \frac{\cot A \cot B - 1}{\cot A + \cot B}$$

$$\equiv \text{LHS.}$$

$$\boxed{3 \sin A - 4 \sin^3 A \equiv \sin 3A}$$

Eyeballing and Mental Gymnastics

1. *3A suggests expansion of "compound angle" twice*
2. *rearrange and simplify*
3. *the RHS is a standard function, easy to expand, twice.*

$$\text{RHS} = \sin 3A$$

$$= \sin(2A + A)$$

$$= \sin 2A \cos A + \cos 2A \sin A$$

$$= 2 \sin A \cos A \cdot \cos A + \sin A (\cos^2 A - \sin^2 A)$$

$$= 2 \sin A \cos^2 A + \sin A \cos^2 A - \sin^3 A$$

$$= 3 \sin A \cos^2 A - \sin^3 A$$

$$= 3 \sin A (1 - \sin^2 A) - \sin^3 A$$

$$= 3 \sin A - 3 \sin^3 A - \sin^3 A$$

$$= 3 \sin A - 4 \sin^3 A$$

$$\equiv \text{LHS.}$$

$$\boxed{\frac{3\tan A - \tan^3 A}{1 - 3\tan^2 A} \equiv \tan 3A}$$

Eyeballing and Mental Gymnastics

1. *RHS* tan 3A *is standard double expansion of* tan(2A + A) *and* tan 2A.
2. *begin with RHS* (*an exception to normal practice*)
3. *rearrange and simplify.*

$$RHS = \tan 3A$$

$$= \tan(2A + A)$$

$$= \frac{\tan 2A + \tan A}{1 - \tan 2A \tan A}$$

$$= \frac{\dfrac{2\tan A}{1 - \tan^2 A} + \tan A}{1 - \dfrac{2\tan A}{1 - \tan^2 A}\tan A}$$

$$= \frac{\dfrac{2\tan A + \tan A - \tan^3 A}{1 - \tan^2 A}}{\dfrac{1 - \tan^2 A - 2\tan^2 A}{1 - \tan^2 A}}$$

$$= \frac{3\tan A - \tan^3 A}{1 - 3\tan^2 A}$$

$$\equiv LHS.$$

$$\tan(45° + A)\tan(45° - A) \equiv \cot(45° + A)\cot(45° - A)$$

Eyeballing and Mental Gymnastics

1. $\tan 45° = 1$, $\cot 45° = 1$
2. *both sides equally complex; therefore maybe easier to work on both sides to reduce to common terms*
3. *( ) suggests expansion*
4. *rearrange and simplify.*

$$\text{LHS} = \tan(45° + A)\tan(45° - A)$$

$$= \frac{(\tan 45° + \tan A)}{(1 - \tan 45° \tan A)} \cdot \frac{(\tan 45° - \tan A)}{(1 + \tan 45° \tan A)}$$

$$= \frac{(1 + \tan A)}{(1 - \tan A)} \cdot \frac{(1 - \tan A)}{(1 + \tan A)}$$

$$= 1$$

$$\text{RHS} = \cot(45° + A)\cot(45° - A)$$

$$= \frac{1}{\tan(45° + A)} \cdot \frac{1}{\tan(45° - A)}$$

$$= \frac{(1 - \tan 45° \tan A)}{(\tan 45° + \tan A)} \cdot \frac{(1 + \tan 45° \tan A)}{(\tan 45° - \tan A)}$$

$$= \frac{(1 - \tan A)}{(1 + \tan A)} \cdot \frac{(1 + \tan A)}{(1 - \tan A)}$$

$$= 1$$

$$\equiv \text{simplified form of LHS.}$$

$$\boxed{\begin{array}{l} \tan 45° = 1 \\ \cot 45° = 1 \end{array}}$$

$$\boxed{\frac{\sin A + \sin 2A}{2 + 3\cos A + \cos 2A} \equiv \tan\frac{A}{2}}$$

Eyeballing and Mental Gymnastics

1. $\sin 2A$, $\cos 2A$ suggest *"double angle" expansion*
2. $(A/2)$ *on RHS suggests "half-angle formula"*
3. *rearrange and simplify.*

$$\text{LHS} = \frac{\sin A + \sin 2A}{2 + 3\cos A + \cos 2A}$$

$$= \frac{\sin A + 2\sin A \cos A}{2 + 3\cos A + (2\cos^2 A - 1)}$$

$$= \frac{\sin A(1 + 2\cos A)}{1 + 3\cos A + 2\cos^2 A}$$

$$= \frac{\sin A(1 + 2\cos A)}{(1 + \cos A)(1 + 2\cos A)}$$

$$= \frac{2\sin\dfrac{A}{2}\cos\dfrac{A}{2}}{1 + \left(2\cos^2\dfrac{A}{2} - 1\right)}$$

$$= \frac{2\sin\dfrac{A}{2}\cos\dfrac{A}{2}}{2\cos^2\dfrac{A}{2}}$$

$$= \frac{\sin\dfrac{A}{2}}{\cos\dfrac{A}{2}}$$

$$= \tan\frac{A}{2}$$

$$\equiv \text{RHS.}$$

$$\boxed{\operatorname{cosec} A \tan \frac{A}{2} - \frac{\cos 2A}{1 + \cos A} \equiv 4 \sin^2 \frac{A}{2}}$$

Eyeballing and Mental Gymnastics

1. $\operatorname{cosec} = 1/\sin$
2. $2A, A/2$ suggest expansion using *"double angle"*, *"half angle" formulas*
3. *common denominator*
4. *rearrange and simplify.*

$$\text{LHS} = \operatorname{cosec} A \tan \frac{A}{2} - \frac{\cos 2A}{1 + \cos A}$$

$$= \frac{1}{2 \sin \frac{A}{2} \cos \frac{A}{2}} \cdot \frac{\sin \frac{A}{2}}{\cos \frac{A}{2}} - \frac{(2 \cos^2 A - 1)}{1 + \cos A}$$

$$= \frac{1}{2 \cos^2 \frac{A}{2}} - \frac{(2 \cos^2 A - 1)}{1 + \left(2 \cos^2 \frac{A}{2} - 1\right)}$$

$$= \frac{1}{2 \cos^2 \frac{A}{2}} - \frac{(2 \cos^2 A - 1)}{2 \cos^2 \frac{A}{2}}$$

$$= \frac{1 - (2 \cos^2 A - 1)}{2 \cos^2 \frac{A}{2}}$$

$$= \frac{2 - 2 \cos^2 A}{2 \cos^2 \frac{A}{2}}$$

$$= \frac{1 - \cos^2 A}{\cos^2 \frac{A}{2}}$$

$$= \frac{\sin^2 A}{\cos^2 \dfrac{A}{2}}$$

$$= \frac{\left(2 \sin \dfrac{A}{2} \cos \dfrac{A}{2} \right)^2}{\cos^2 \dfrac{A}{2}}$$

$$= \frac{4 \sin^2 \dfrac{A}{2} \cos^2 \dfrac{A}{2}}{\cos^2 \dfrac{A}{2}}$$

$$= 4 \sin^2 \frac{A}{2}$$

$$\equiv \text{RHS.}$$

$$\frac{\sin 2A \cos A - 2 \cos 2A \sin A}{2 \sin A - \sin 2A} \equiv 2 \cos^2 \frac{A}{2}$$

Eyeballing and Mental Gymnastics

1. $2A$ suggests "double angle" expansion
2. $2 \cos^2(A/2)$ on RHS suggest $\cos A \equiv 2 \cos^2(A/2) - 1$
3. rearrange and simplify.

$$\text{LHS} = \frac{\sin 2A \cos A - 2 \cos 2A \sin A}{2 \sin A - \sin 2A}$$

$$= \frac{2 \sin A \cos A \cdot \cos A - 2 \cos 2A \sin A}{2 \sin A - 2 \sin A \cos A}$$

$$= \frac{2 \sin A(\cos^2 A - \cos 2A)}{2 \sin A(1 - \cos A)}$$

$$= \frac{\cos^2 A - (2 \cos^2 A - 1)}{1 - \cos A}$$

$$= \frac{1 - \cos^2 A}{1 - \cos A}$$

$$= \frac{(1 - \cos A)(1 + \cos A)}{(1 - \cos A)}$$

$$= (1 + \cos A)$$

$$= 1 + \left(2 \cos^2 \frac{A}{2} - 1\right)$$

$$= 2 \cos^2 \frac{A}{2}$$

$$\equiv \text{RHS}.$$

$$\boxed{\frac{\cos A}{1 + \cos 2A} + \frac{\sin A}{1 - \cos 2A} \equiv \frac{\sin A + \cos A}{\sin 2A}}$$

Eyeballing and Mental Gymnastics

1. $2A$ suggests expansion of *"compound angles"*
2. *rearrange and simplify.*

$$\text{LHS} = \frac{\cos A}{1 + \cos 2A} + \frac{\sin A}{1 - \cos 2A}$$

$$= \frac{\cos A(1 - \cos 2A) + \sin A(1 + \cos 2A)}{(1 + \cos 2A)(1 - \cos 2A)}$$

$$= \frac{\cos A(1 - (2\cos^2 A - 1)) + \sin A(2\cos^2 A)}{(1 - \cos^2 2A)}$$

$$= \frac{\cos A(2 - 2\cos^2 A) + 2\sin A \cos^2 A}{\sin^2 2A}$$

$$= \frac{2\cos A - 2\cos^3 A + 2\sin A \cos^2 A}{\sin^2 2A}$$

$$= \frac{2\cos A(1 - \cos^2 A + \sin A \cos A)}{\sin^2 2A}$$

$$= \frac{2\cos A(\sin^2 A + \sin A \cos A)}{\sin^2 2A}$$

$$= \frac{2\cos A \sin A(\sin A + \cos A)}{\sin^2 2A}$$

$$= \frac{\sin 2A(\sin A + \cos A)}{\sin^2 2A}$$

$$= \frac{(\sin A + \cos A)}{\sin 2A}$$

$$\equiv \text{RHS}.$$

$$\boxed{\frac{1+\tan A}{1-\tan A} + \frac{1-\tan A}{1+\tan A} \equiv 2\sec 2A}$$

Eyeballing and Mental Gymnastics

1. $t = s/c$, $\sec = 1/\cos$
2. $2A$ suggests "double angle"
3. *common denominator*
4. *rearrange and simplify.*

$$\begin{aligned}
\text{LHS} &= \frac{1+\tan A}{1-\tan A} + \frac{1-\tan A}{1+\tan A} \\[2mm]
&= \frac{(1+\tan A)^2 + (1-\tan A)^2}{(1-\tan A)(1+\tan A)} \\[2mm]
&= \frac{(1+2\tan A+\tan^2 A) + (1-2\tan A+\tan^2 A)}{1-\tan^2 A} \\[2mm]
&= \frac{2+2\tan^2 A}{1-\tan^2 A} \\[2mm]
&= \frac{2(1+\tan^2 A)}{1-\tan^2 A} \\[2mm]
&= 2 \cdot (\sec^2 A)\frac{1}{1 - \dfrac{\sin^2 A}{\cos^2 A}} \\[2mm]
&= 2 \cdot \sec^2 A \cdot \frac{\cos^2 A}{(\cos^2 A - \sin^2 A)} \\[2mm]
&= 2\frac{1}{\cos 2A} \\[2mm]
&= 2\sec 2A \\[2mm]
&\equiv \text{RHS.}
\end{aligned}$$

$$\boxed{\frac{\sin 2A + \cos 2A + 1}{\sin 2A + \cos 2A - 1} \equiv \frac{\tan(45° + A)}{\tan A}}$$

Eyeballing and Mental Gymnastics

1. *$2A$ suggests expansion of "compound angle"*
2. *$\tan(45° + A)$ suggests expansion of "compound angle"*
3. *both sides to be simplified before identity is established*
4. *$t = s/c$*
5. *rearrange and simplify.*

$$\text{LHS} = \frac{\sin 2A + \cos 2A + 1}{\sin 2A + \cos 2A - 1}$$

$$= \frac{(2 \sin A \cos A) + (\cos^2 A - \sin^2 A) + (\cos^2 A + \sin^2 A)}{(2 \sin A \cos A) + (\cos^2 A - \sin^2 A) - (\cos^2 A + \sin^2 A)}$$

$$= \frac{2 \sin A \cos A + 2 \cos^2 A}{2 \sin A \cos A - 2 \sin^2 A}$$

$$= \frac{2 \cos A (\sin A + \cos A)}{2 \sin A (\cos A - \sin A)}$$

$$= \frac{1}{\tan A} \left(\frac{\tan A + 1}{1 - \tan A} \right) \qquad \left|\begin{array}{l}\text{dividing all} \\ \text{terms by } \cos A.\end{array}\right.$$

$$\text{RHS} = \frac{\tan(45° + A)}{\tan A}$$

$$= \left(\frac{\tan 45° + \tan A}{1 - \tan 45° \tan A} \right) \left(\frac{1}{\tan A} \right)$$

$$= \frac{1}{\tan A} \left(\frac{1 + \tan A}{1 - \tan A} \right) \qquad |\tan 45° = 1$$

$$\equiv \text{simplified form of LHS}$$

$$\sqrt{\left(\frac{1 - \sin A}{1 + \sin A}\right)} \equiv \sec A - \tan A$$

Eyeballing and Mental Gymnastics

1. *square the sq root function*
2. *rearrange and simplify.*

$$\text{LHS} = \sqrt{\frac{1 - \sin A}{1 + \sin A}}$$

$$(\text{LHS})^2 = \frac{1 - \sin A}{1 + \sin A}$$

$$= \frac{1 - \sin A}{1 + \sin A} \cdot \left(\frac{1 - \sin A}{1 - \sin A}\right)$$

$$= \frac{(1 - \sin A)^2}{1 - \sin^2 A}$$

$$= \frac{(1 - \sin A)^2}{\cos^2 A}$$

$$= \left(\frac{1}{\cos A} - \frac{\sin A}{\cos A}\right)^2$$

$$= (\sec A - \tan A)^2$$

$$\equiv (\text{RHS})^2$$

$$\therefore \text{LHS} \equiv \text{RHS}.$$

$$\boxed{\frac{\sin^2 2A + 2\cos 2A - 1}{\sin^2 2A + 3\cos 2A - 3} \equiv \frac{1}{1 - \sec 2A}}$$

Eyeballing and Mental Gymnastics

1. s^2 suggests $s^2 + c^2 \equiv 1$
2. *factorisation of LHS*
3. *rearrange and simplify.*

$$\text{LHS} = \frac{\sin^2 2A + 2\cos 2A - 1}{\sin^2 2A + 3\cos 2A - 3}$$

$$= \frac{(1 - \cos^2 2A) + 2\cos 2A - 1}{(1 - \cos^2 2A) + 3\cos 2A - 3}$$

$$= \frac{\cos 2A(2 - \cos 2A)}{3\cos 2A - \cos^2 2A - 2}$$

$$= \frac{\cos 2A(2 - \cos 2A)}{(\cos 2A - 1)(2 - \cos 2A)}$$

$$= \frac{\cos 2A}{(\cos 2A - 1)} \qquad \begin{array}{l}\text{divide numerator} \\ \text{and denominator} \\ \text{by } \cos 2A\end{array}$$

$$= \frac{1}{1 - \sec 2A}$$

$$\equiv \text{RHS}.$$

$$\boxed{\frac{\sin(A+45°)}{\cos(A+45°)} + \frac{\cos(A+45°)}{\sin(A+45°)} \equiv 2\sec 2A}$$

Eyeballing and Mental Gymnastics

1. $(A+45°)$ *suggests expansion of "compound angle"*
2. *2A suggests "double angle"*
3. $\sec = 1/\cos$
4. *common denominator*
5. *rearrange and simplify.*

$$\text{LHS} = \frac{\sin(A+45°)}{\cos(A+45°)} + \frac{\cos(A+45°)}{\sin(A+45°)}$$

$$= \frac{\sin^2(A+45°) + \cos^2(A+45°)}{\cos(A+45°)\sin(A+45°)}$$

$$= \frac{1}{\cos(A+45°)\sin(A+45°)}$$

$$= \frac{2}{2\cos(A+45°)\sin(A+45°)}$$

$$= \frac{2}{\sin 2(A+45°)}$$

$$= \frac{2}{\sin(2A+90°)}$$

$$= \frac{2}{\cos 2A} \qquad \left| \begin{array}{l} \text{since } \sin(x+90) \\ \quad = \cos x \end{array} \right.$$

$$= 2\sec 2A$$

$$\equiv \text{RHS}.$$

(turns out that the problem is easier than expected, as the use of common denominator resulted in $s^2 + c^2 \equiv 1$; hence there is no need to expand $(A+45°)$!)

$$\tan A + \tan 2A \equiv \frac{\sin A(4\cos^2 A - 1)}{\cos A \cos 2A}$$

Eyeballing and Mental Gymnastics

1. $t = s/c$
2. $2A$ suggests expansion of "double angle"
3. *rearrange and simplify.*

$$\text{LHS} = \tan A + \tan 2A$$

$$= \frac{\sin A}{\cos A} + \frac{\sin 2A}{\cos 2A}$$

$$= \frac{\sin A \cos 2A + \sin 2A \cos A}{\cos A \cos 2A}$$

$$= \frac{\sin A(2\cos^2 A - 1) + \cos A(2\sin A \cos A)}{\cos A \cos 2A}$$

$$= \frac{\sin A(2\cos^2 A - 1 + 2\cos^2 A)}{\cos A \cos 2A}$$

$$= \frac{\sin A(4\cos^2 A - 1)}{\cos A \cos 2A}$$

$$\equiv \text{RHS}.$$

$$(\tan A - \operatorname{cosec} A)^2 - (\cot A - \sec A)^2 \equiv 2(\operatorname{cosec} A - \sec A)$$

Eyeballing and Mental Gymnastics

1. $t = s/c$, $\operatorname{cosec} = 1/\sin$, $\cot = c/s$, $\sec = 1/\cos$
2. $(\ \ )^2$ *suggests* $s^2 + c^2 \equiv 1$
3. *rearrange and simplify.*

$$\text{LHS} = (\tan A - \operatorname{cosec} A)^2 - (\cot A - \sec A)^2$$

$$= (\tan^2 A - 2\tan A \operatorname{cosec} A + \operatorname{cosec}^2 A)$$

$$- (\cot^2 A - 2\cot A \sec A + \sec^2 A)$$

$$= (\tan^2 A - \sec^2 A) + (\operatorname{cosec}^2 A - \cot^2 A)$$

$$- 2\frac{\sin A}{\cos A} \cdot \frac{1}{\sin A} + 2\frac{\cos A}{\sin A} \cdot \frac{1}{\cos A}$$

$$= (-1) + (1) - \frac{2}{\cos A} + \frac{2}{\sin A}$$

$$= 2\left(\frac{1}{\sin A} - \frac{1}{\cos A}\right)$$

$$= 2(\operatorname{cosec} A - \sec A)$$

$$\equiv \text{RHS.}$$

$$2 \sin 2A(1 - 2 \sin^2 A) \equiv \sin 4A$$

Eyeballing and Mental Gymnastics

1. $\sin 4A$ *suggests* $\sin 2(2A)$
2. $(1 - 2 \sin^2 A)$ *equals* $\cos 2A$
3. *rearrange and simplify.*

$$\text{LHS} = 2 \sin 2A(1 - 2 \sin^2 A)$$

$$= 2 \sin 2A \cos 2A$$

$$= \sin 4A$$

$$\equiv \text{RHS.}$$

Alternatively since $\sin 4A$ is a standard expression we can proceed with the RHS.

$$\text{RHS} = \sin 4A$$

$$= 2 \sin 2A \cos 2A$$

$$= 2 \sin 2A(1 - 2 \sin^2 A)$$

$$\equiv \text{LHS.}$$

$$\boxed{\cos 3A \equiv 4 \cos^3 A - 3 \cos A}$$

Eyeballing and Mental Gymnastics

1. *3A suggests double expansion of "compound angle"*
2. *rearrange and simplify.*

$$\begin{aligned}
\text{LHS} &= \cos 3A \\
&= \cos(2A + A) \\
&= \cos 2A \cos A - \sin 2A \sin A \\
&= \cos A(\cos^2 A - \sin^2 A) - \sin A(2 \sin A \cos A) \\
&= \cos^3 A - \sin^2 A \cos A - 2 \sin^2 A \cos A \\
&= \cos^3 A - 3 \sin^2 A \cos A \\
&= \cos^3 A - 3 \cos A(1 - \cos^2 A) \\
&= \cos^3 A - 3 \cos A + 3 \cos^3 A \\
&= 4 \cos^3 A - 3 \cos A \\
&\equiv \text{RHS.}
\end{aligned}$$

$$\boxed{32 \cos^6 A - 48 \cos^4 A + 18 \cos^2 A - 1 \equiv \cos 6A}$$

Eyeballing and Mental Gymnastics

1. $\cos 6A$ on RHS suggest "double angle" formula for $(3A)$
2. $3A$ suggests expansion of "triple angle"
3. this is one of the rare occasions where it may be easier to start with the simpler RHS
4. rearrange and simplify.

$\text{RHS} = \cos 6A$

$= \cos 2(3A)$

$= 2\cos^2(3A) - 1$

$= 2(4\cos^3 A - 3\cos A)^2 - 1$ | see previous proof

$= 2\cos^2 A(4\cos^2 A - 3)^2 - 1$

$= 2\cos^2 A(16\cos^4 A + 24\cos^2 A + 9) - 1$

$= 32\cos^6 A + 48\cos^4 A + 18\cos^2 A - 1$

$\equiv \text{LHS.}$

$$\boxed{\cos 4A + 4\cos 2A + 3 \equiv 8\cos^4 A}$$

Eyeballing and Mental Gymnastics

1. $\cos 4A$, $\cos 2A$ *suggest expansion of "double angle"*
2. *rearrange and simplify.*

$$\text{LHS} = (\cos 4A) + 4\cos 2A + 3$$

$$= (2\cos^2 2A - 1) + 4\cos 2A + 3$$

$$= 2\cos^2 2A + 4\cos 2A + 2$$

$$= 2(\cos^2 2A + 2\cos 2A + 1)$$

$$= 2(\cos 2A + 1)^2$$

$$= 2(2\cos^2 A - 1 + 1)^2$$

$$= 2(4\cos^4 A)$$

$$= 8\cos^4 A$$

$$\equiv \text{RHS}.$$

$$\boxed{\frac{\sin 3A}{\sin A} - \frac{\cos 3A}{\cos A} \equiv 2}$$

Eyeballing and Mental Gymnastics

1. $\sin 3A$, $\cos 3A$ *suggest expansion of "compound angle"*
2. *rearrange and simplify.*

$$\text{LHS} = \frac{\sin 3A}{\sin A} - \frac{\cos 3A}{\cos A}$$

$$= \frac{(3 \sin A - 4 \sin^3 A)}{\sin A} - \frac{4(\cos^3 A - 3 \cos A)}{\cos A} \qquad \begin{array}{l} \text{see proofs} \\ \text{on p. 324} \\ \text{and 340} \end{array}$$

$$= \frac{\sin A (3 - 4 \sin^2 A)}{\sin A} - \frac{\cos A (4 \cos^2 A - 3)}{\cos A}$$

$$= 3 - 4 \sin^2 A - 4 \cos^2 A + 3$$

$$= 6 - 4(\sin^2 A + \cos^2 A)$$

$$= 6 - 4$$

$$= 2$$

$$\equiv \text{RHS.}$$

$$\boxed{\sin^4 A + \cos^4 A \equiv \frac{3}{4} + \frac{1}{4}\cos 4A}$$

Eyeballing and Mental Gymnastics

1. *The* cos 4A *on the RHS, a standard function, suggests that it may be easier to start with the RHS through a double expansion of* cos 2(2A)
2. *rearrange and simplify.*

$$\text{RHS} = \frac{3}{4} + \frac{1}{4}\cos 4A$$

$$= \frac{3}{4} + \frac{1}{4}(2\cos^2 2A - 1)$$

$$= \frac{3}{4} + \frac{1}{2}\cos^2 2A - \frac{1}{4}$$

$$= \frac{1}{2} + \frac{1}{2}(2\cos^2 A - 1)^2$$

$$= \frac{1}{2} + \frac{1}{2}(4\cos^4 A - 4\cos^2 A + 1)$$

$$= 1 + 2\cos^4 A - 2\cos^2 A$$

$$= \cos^4 A + \cos^4 A - 2\cos^2 A + 1$$

$$= \cos^4 A + (\cos^2 A - 1)^2$$

$$= \cos^4 A + (\sin^2 A)^2$$

$$= \cos^4 A + \sin^4 A$$

$$\equiv \text{LHS.}$$

$$8 \cos^4 A - 4 \cos 2A - 3 \equiv \cos 4A$$

Eyeballing and Mental Gymnastics

1. *cos 4A on the RHS is a standard expression and suggests the expansion of* $\cos 4A \to \cos 2A \to \cos A$
2. *hence, easier to begin from the RHS*
3. *rearrange and simplify.*

$$\begin{aligned}
\text{RHS} &= \cos 4A \\[6pt]
&= \cos 2(2A) \\[6pt]
&= 2 \cos^2(2A) - 1 \\[6pt]
&= 2(2 \cos^2 A - 1)^2 - 1 \\[6pt]
&= 2(4 \cos^4 A - 4 \cos^2 A + 1) - 1 \\[6pt]
&= 8 \cos^4 A - 8 \cos^2 A + 2 - 1 \\[6pt]
&= 8 \cos^4 A - 4(2 \cos^2 A - 1) - 2 - 1 \\[6pt]
&= 8 \cos^4 A - 4 \cos 2A - 3 \\[6pt]
&\equiv \text{LHS.}
\end{aligned}$$

$$\boxed{1 - 8\sin^2 A \cos^2 A \equiv \cos 4A}$$

Eyeballing and Mental Gymnastics

1. $\cos 4A$ — *standard expression; therefore easier to begin with RHS and expand twice* $\cos 4A \to \cos 2A \to \cos A$
2. s^2, c^2 *suggest* $s^2 + c^2 \equiv 1$
3. *rearrange and simplify.*

$$\text{RHS} = \cos 4A$$

$$= 2\cos^2(2A) - 1$$

$$= 2(2\cos^2 A - 1)^2 - 1$$

$$= 2(4\cos^4 A - 4\cos^2 A + 1) - 1$$

$$= (8\cos^4 A - 8\cos^2 A + 2) - 1$$

$$= 8\cos^2 A(\cos^2 A - 1) + 1$$

$$= 8\cos^2 A(-\sin^2 A) + 1$$

$$= 1 - 8\sin^2 A \cos^2 A$$

$$\equiv \text{LHS.}$$

$$\boxed{\frac{1 - \cos 2A + \sin A}{\sin 2A + \cos A} \equiv \tan A}$$

Eyeballing and Mental Gymnastics

1. $2A$ suggests expansion of *"double angle"*
2. $t = s/c$
3. *rearrange and simplify.*

$$\text{LHS} = \frac{1 - \cos 2A + \sin A}{\sin 2A + \cos A}$$

$$= \frac{1 - (\cos^2 A - \sin^2 A) + \sin A}{2 \sin A \cos A + \cos A}$$

$$= \frac{(\cos^2 A + \sin^2 A) - (\cos^2 A - \sin^2 A) + \sin A}{\cos A (2 \sin A + 1)}$$

$$= \frac{2 \sin^2 A + \sin A}{\cos A (2 \sin A + 1)}$$

$$= \frac{\sin A \, (2 \sin A + 1)}{\cos A \, (2 \sin A + 1)}$$

$$= \tan A$$

$$\equiv \text{LHS}.$$

Level-Three-Games

Not-So-Easy Proofs

Angles in a Triangle

$$(A+B+C) = 180°$$

$$\sin A + \sin B + \sin C \equiv 4 \cos \frac{A}{2} \cos \frac{B}{2} \cos \frac{C}{2}$$

Eyeballing and Mental Gymnastics

1. $(\sin A + \sin B)$ *suggests* $\sin S + \sin T$ *formula*
2. $C = 180° - (A + B)$
3. $A/2, B/2, C/2$ *suggest "half angle" formula*
4. *rearrange and simplify.*

see next page

LHS $= \sin A + \sin B + \sin C$

$\quad = 2 \sin \dfrac{A+B}{2} \cos \dfrac{A-B}{2} + \sin C$

$\quad = 2 \sin \left(\dfrac{180° - C}{2} \right) \cos \dfrac{(A-B)}{2} + 2 \sin \dfrac{C}{2} \cos \dfrac{C}{2}$

$\quad = 2 \cos \dfrac{C}{2} \cos \dfrac{(A-B)}{2} + 2 \sin \dfrac{C}{2} \cos \dfrac{C}{2} \qquad \left| \; \sin \left(90° - \dfrac{C}{2} \right) = \cos \dfrac{C}{2} \right.$

$\quad = 2 \cos \dfrac{C}{2} \left(\cos \dfrac{A-B}{2} + \sin \dfrac{C}{2} \right)$

$\quad = 2 \cos \dfrac{C}{2} \left(\cos \dfrac{A-B}{2} + \cos \dfrac{A+B}{2} \right) \qquad \left| \; \sin \dfrac{C}{2} = \cos \left(90° - \dfrac{C}{2} \right) = \cos \dfrac{A+B}{2} \right.$

$\quad = 2 \cos \dfrac{C}{2} \left(2 \cos \dfrac{(A-B)+(A+B)}{4} \cos \dfrac{(A-B)-(A+B)}{4} \right)$

$\quad = 2 \cos \dfrac{C}{2} \left(2 \cos \dfrac{2A}{4} \cos \dfrac{(-2B)}{4} \right)$

$\quad = 4 \cos \dfrac{C}{2} \cos \dfrac{A}{2} \cos \dfrac{B}{2} \qquad\qquad | \cos(-B) = \cos B$

$\quad = 4 \cos \dfrac{A}{2} \cos \dfrac{B}{2} \cos \dfrac{C}{2}$

$\quad \equiv$ RHS.

$$\sin(A+B+C) \left.\vphantom{\begin{matrix}\\\\\\\\\end{matrix}}\right\} \equiv \left\{ \begin{matrix} \sin A \cos B \cos C \\ + \sin B \cos C \cos A \\ + \sin C \cos A \cos B \\ - \sin A \sin B \sin C \end{matrix} \right.$$

Eyeballing and Mental Gymnastics

1. $(A+B+C)$ suggests expansion of "compound angle" twice
2. rearrange and simplify.

$$\text{LHS} = \sin(A+B+C)$$

$$= \sin(A+B)\cos C + \cos(A+B)\sin C$$

$$= \cos C(\sin A \cos B + \cos A \sin B)$$

$$+ \sin C(\cos A \cos B - \sin A \sin B)$$

$$= \sin A \cos B \cos C + \sin B \cos C \cos A$$

$$+ \sin C \cos A \cos B - \sin A \sin B \sin C$$

$$\equiv \text{RHS}.$$

---

*Note that:

this general identity is for any three angles A, B and C; they need not necessarily be for the three angles of a triangle.

Angles in a Triangle

$$(A + B + C = 180°)$$

$$\sin A \cos B \cos C + \sin B \cos C \cos A + \sin C \cos A \cos B$$

$$\equiv \sin A \sin B \sin C$$

Eyeballing and Mental Gymnastics

1. *This identity is easiest proved as a follow-up from the previous proof of the general identity:*

$$\sin(A + B + C) \equiv \begin{cases} \sin A \cos B \cos C \\ + \sin B \cos C \cos A \\ + \sin C \cos A \cos B \\ - \sin A \sin B \sin C \end{cases}$$

2. $\sin(A + B + C) = \sin 180° = 0.$

Since $\sin(A + B + C) = \sin 180° = 0,$

$$\therefore \quad \left. \begin{array}{l} \sin A \cos B \cos C \\ + \sin B \cos C \cos A \\ + \sin C \cos A \cos B \\ - \sin A \sin B \sin C \end{array} \right\} = 0$$

$$\therefore \quad \left. \begin{array}{l} \sin A \cos B \cos C \\ + \sin B \cos C \cos A \\ + \sin C \cos A \cos B \end{array} \right\} = \sin A \sin B \sin C$$

$$\therefore \text{LHS} \equiv \text{RHS}.$$

(This identity is extremely valuable for proving many subsequent identities involving the three angles of a triangle. Without knowledge of the preceding general identity, this identity is more difficult to prove. Try it out for yourself)

Angles in a Triangle

$$(A+B+C) = 180°$$

$$\boxed{\sin 2A + \sin 2B + \sin 2C \equiv 4 \sin A \sin B \sin C}$$

Eyeballing and Mental Gymnastics

1. $\sin 2A + \sin 2B$ *suggests* $\sin S + \sin T$ *formula*
2. $\sin 2C$ *suggests "double angle" formula*
3. $(A+B+C) = 180°$ *therefore express values for* $(180° - C)$ *in terms of* $A+B$
4. *rearrange and simplify.*

$$\text{LHS} = \sin 2A + \sin 2B + \sin 2C$$

$$= 2 \sin \left(\frac{2A+2B}{2} \right) \cos \left(\frac{2A-2B}{2} \right) + 2 \sin C \cos C$$

$$= 2 \sin(A+B) \cos(A-B) + 2 \sin C \cos C$$

$$= 2 \sin C \cos(A-B) - 2 \sin C \cos(A+B) \qquad \left| \begin{array}{l} \text{see identities} \\ \text{below} \end{array} \right.$$

$$= 2 \sin C (\cos(A-B) - \cos(A+B))$$

$$= 2 \sin C \left(-2 \sin \frac{(A-B)+(A+B)}{2} \sin \frac{(A-B)-(A+B)}{2} \right)$$

$$= 2 \sin C (-2 \sin A \sin(-B))$$

$$= 4 \sin C \sin A \sin B$$

$$= 4 \sin A \sin B \sin C$$

$$\equiv \text{RHS.}$$

$$\begin{array}{ll} \sin(A+B) = \sin(180° - (A+B)) & \cos C = -\cos(180° - C) \\ \qquad\qquad = \sin C & \qquad\quad = -\cos(A+B) \end{array}$$

Angles in a Triangle

$$(A+B+C) = 180°$$

$$\boxed{\sin 2A + \sin 2B + \sin 2C \equiv 4 \sin A \sin B \sin C}$$

For this beautiful identity let's explore a second approach.

Eyeballing and Mental Gymnastics

1. $2A$, $2B$, $2C$ suggest expansion of *"double angles"*
2. $A+B+C = 180°$
3. *rearrange and simplify.*

$$\text{LHS} = \sin 2A + \sin 2B + \sin 2C$$

$$= \quad 2 \sin A \cos A$$
$$+ 2 \sin B \cos B$$
$$+ 2 \sin C \cos C$$

$$= \quad 2 \sin A(\sin B \sin C - \cos B \cos C)$$
$$+ 2 \sin B(\sin C \sin A - \cos C \cos A) \qquad \left|\begin{array}{l}\text{see identities}\\ \text{below}\end{array}\right.$$
$$+ 2 \sin C(\sin A \sin B - \cos A \cos B)$$

$$= 6 \sin A \sin B \sin C - 2 \sin A \cos B \cos C$$
$$- 2 \sin B \cos C \cos A$$
$$- 2 \sin C \cos A \cos B$$

$$= 6 \sin A \sin B \sin C - 2(\sin A \sin B \sin C) \qquad \left|\begin{array}{l}\text{from previous}\\ \text{proof on p. 353}\end{array}\right.$$
$$= 4 \sin A \sin B \sin C$$

$$\equiv \text{RHS}.$$

$$\begin{aligned}\cos A &= \cos[180° - (B+C)] \\ &= -\cos(B+C) \\ &= \sin B \sin C - \cos B \cos C\end{aligned} \qquad \left| A = 180° - (B+C)\right.$$

$$\left.\begin{array}{l}\sin A \cos B \cos C \\ \sin B \cos C \cos A \\ \sin C \cos A \cos B\end{array}\right\} = \begin{array}{l}\sin(A+B+C) \\ + \sin A \sin B \sin C\end{array} \qquad \left|\begin{array}{l}\sin(A+B+C) \\ = \sin(180°) \\ = 0\end{array}\right.$$

Angles in a Triangle

$$(A + B + C) = 180°$$

$$\boxed{\tan A + \tan B + \tan C \equiv \tan A \tan B \tan C}$$

Eyeballing and Mental Gymnastics

1. $t = s/c$
2. $(A + B + C) = 180°$
3. *rearrange and simplify.*

$$LHS = \tan A + \tan B + \tan C$$

$$= \frac{\sin A}{\cos A} + \frac{\sin B}{\cos B} + \frac{\sin C}{\cos C}$$

$$= \frac{\sin A \cos B \cos C + \sin B \cos C \cos A + \sin C \cos A \cos B}{\cos A \cos B \cos C}$$

$$= \frac{\sin A \sin B \sin C}{\cos A \cos B \cos C} \qquad \begin{vmatrix} \text{see identity*} \\ \text{below} \end{vmatrix}$$

$$= \tan A \tan B \tan C$$

$$\equiv RHS.$$

---

*For $(A + B + C) = 180°$

$$\left. \begin{array}{l} \sin A \cos B \cos C \\ + \sin B \cos C \cos A \\ + \sin C \cos A \cos B \end{array} \right\} = \sin A \sin B \sin C \qquad \begin{vmatrix} \text{from previous} \\ \text{proof on p. 353} \end{vmatrix}$$

An alternative proof is to begin with $\tan C = -\tan(A + B)$ with expansion of $\tan(A + B)$ so that $\tan A + \tan B$ can be expressed in terms of $\tan A$, $\tan B$ and $\tan C$. Try it.

Angles in a Triangle

$$(A + B + C) = 180°$$

$$\cos A + \cos B + \cos C \equiv 4 \sin \frac{A}{2} \sin \frac{B}{2} \sin \frac{C}{2} + 1$$

Eyeballing and Mental Gymnastics

1. $\cos A + \cos B$ *suggests* $\cos S + \cos T$ *formula*
2. *since* $(A + B + C) = 180°$, *then* $C/2$ *is complementary angle of* $(A + B)/2$;
   i.e. $(90 - (A + B)/2)$
3. *rearrange and simplify.*

$$\text{LHS} = (\cos A + \cos B) + (\cos C)$$

$$= \left(2 \cos \frac{A+B}{2} \cos \frac{A-B}{2} \right) + \left(1 - 2 \sin^2 \frac{C}{2} \right)$$

$$= 2 \sin \frac{C}{2} \cos \frac{A-B}{2} + 1 - 2 \sin^2 \frac{C}{2}$$

$$= 2 \sin \frac{C}{2} \left(\cos \frac{A-B}{2} - \sin \frac{C}{2} \right) + 1$$

$$= 2 \sin \frac{C}{2} \left(\cos \frac{A-B}{2} - \cos \frac{A+B}{2} \right) + 1$$

$$= 2 \sin \frac{C}{2} \left(-2 \sin \frac{1}{2} \left(\frac{A-B}{2} + \frac{A+B}{2} \right) \sin \frac{1}{2} \left(\frac{A-B}{2} - \frac{A+B}{2} \right) \right) + 1$$

$$= 2 \sin \frac{C}{2} \left(-2 \sin \frac{A}{2} \sin \left(\frac{-B}{2} \right) \right) + 1$$

$$= 4 \sin \frac{C}{2} \sin \frac{A}{2} \sin \frac{B}{2} + 1$$

$$= 4 \sin \frac{A}{2} \sin \frac{B}{2} \sin \frac{C}{2} + 1$$

$$\equiv \text{RHS}.$$

$$\cos 3A + \cos 2A \equiv 2 \cos \frac{5A}{2} \cos \frac{A}{2}$$

Eyeballing and Mental Gymnastics

1. $\cos S + \cos T \equiv 2 \cos (S+T)/2 \cos (S-T)/2$
2. *rearrange and simplify.*

$$\text{LHS} = \cos 3A + \cos 2A$$

$$= 2 \cos \frac{3A+2A}{2} \cos \frac{3A-2A}{2}$$

$$= 2 \cos \frac{5A}{2} \cos \frac{A}{2}$$

$$\equiv \text{RHS.}$$

$$\boxed{\sin 5A - \sin 3A \equiv 2 \sin A \cos 4A}$$

Eyeballing and Mental Gymnastics

1. $\sin S - \sin T = 2 \cos (S+T)/2 \sin (S-T)/2$
2. *rearrange and simplify.*

$$\text{LHS} = \sin 5A + \sin 3A$$

$$= 2 \cos \frac{5A+3A}{2} \sin \frac{5A-3A}{2}$$

$$= 2 \cos 4A \sin A$$

$$= 2 \sin A \cos 4A$$

$$\equiv \text{RHS}.$$

$$\frac{\sin A + \sin B}{\sin A - \sin B} \equiv \tan \frac{A+B}{2} \cot \frac{A-B}{2}$$

Eyeballing and Mental Gymnastics

1. $\sin S + \sin T \equiv 2 \sin (S+T)/2 \cos (S-T)/2$
2. $\sin S - \sin T \equiv 2 \cos (S+T)/2 \cos (S-T)/2$
3. *rearrange and simplify.*

$$\text{LHS} = \frac{\sin A + \sin B}{\sin A - \sin B}$$

$$= \frac{2 \sin \dfrac{A+B}{2} \cos \dfrac{A-B}{2}}{2 \cos \dfrac{A+B}{2} \sin \dfrac{A-B}{2}}$$

$$= \tan \frac{A+B}{2} \cot \frac{A-B}{2}$$

$$\equiv \text{RHS.}$$

$$\boxed{\frac{\sin(2A+B)+\sin B}{\cos(2A+B)+\cos B} \equiv \tan(A+B)}$$

Eyeballing and Mental Gymnastics

1. $\dfrac{\sin S + \sin T}{\cos S + \cos T}$
2. *rearrange and simplify.*

$$\text{LHS} = \frac{\sin(2A+B)+\sin B}{\cos(2A+B)+\cos B}$$

$$= \frac{2\sin\dfrac{(2A+B)+B}{2}\cos\dfrac{(2A+B)-B}{2}}{2\cos\dfrac{(2A+B)+B}{2}\cos\dfrac{(2A+B)-B}{2}}$$

$$= \frac{2\sin(A+B)\cos A}{2\cos(A+B)\cos A}$$

$$= \tan(A+B)$$

$$\equiv \text{RHS.}$$

$$\boxed{\frac{\cos A + \cos B}{\cos A - \cos B} \equiv - \cot \frac{A+B}{2} \cot \frac{A-B}{2}}$$

Eyeballing and Mental Gymnastics

1. $\cos S + \cos T$
2. $\cos S - \cos T$
3. *rearrange and simplify.*

$$\text{LHS} = \frac{\cos A + \cos B}{\cos A - \cos B}$$

$$= \frac{2 \cos \dfrac{A+B}{2} \cos \dfrac{A-B}{2}}{-2 \sin \dfrac{A+B}{2} \sin \dfrac{A-B}{2}}$$

$$= - \cot \frac{A+B}{2} \cot \frac{A-B}{2}$$

$$\equiv \text{RHS.}$$

$$\boxed{\frac{\sin A + \sin B}{\cos A + \cos B} \equiv \tan \frac{A+B}{2}}$$

Eyeballing and Mental Gymnastics

1. $\sin S + \sin T$
2. $\cos S + \cos T$
3. *rearrange and simplify.*

$$\text{LHS} = \frac{\sin A + \sin B}{\cos A + \cos B}$$

$$= \frac{2 \sin \dfrac{A+B}{2} \cos \dfrac{A-B}{2}}{2 \cos \dfrac{A+B}{2} \cos \dfrac{A-B}{2}}$$

$$= \frac{\sin \dfrac{A+B}{2}}{\cos \dfrac{A+B}{2}}$$

$$= \tan \frac{A+B}{2}$$

$$\equiv \text{RHS.}$$

$$\boxed{\frac{\sin A - \sin B}{\cos A - \cos B} \equiv -\cot \frac{A+B}{2}}$$

Eyeballing and Mental Gymnastics

1. $\sin S - \sin T$
2. $\cos S - \cos T$
3. *rearrange and simplify.*

$$\text{LHS} = \frac{\sin A - \sin B}{\cos A - \cos B}$$

$$= \frac{2 \cos \dfrac{A+B}{2} \sin \dfrac{A-B}{2}}{-2 \sin \dfrac{A+B}{2} \sin \dfrac{A-B}{2}}$$

$$= -\frac{\cos \dfrac{A+B}{2}}{\sin \dfrac{A+B}{2}}$$

$$= -\cot \frac{A+B}{2}$$

$$\equiv \text{RHS.}$$

$$\boxed{\frac{\sin A + \sin 3A}{2 \sin 2A} \equiv \cos A}$$

Eyeballing and Mental Gymnastics

1. $\sin S + \sin T$
2. *rearrange and simplify.*

$$
\begin{aligned}
\text{LHS} &= \frac{\sin A + \sin 3A}{2 \sin 2A} \\
&= \frac{2 \sin \dfrac{A + 3A}{2} \cos \dfrac{A - 3A}{2}}{2 \sin 2A} \\
&= \frac{2 \sin 2A \cos(-A)}{2 \sin 2A} \\
&= \cos(-A) \\
&= \cos A \\
&\equiv \text{RHS.}
\end{aligned}
$$

$$\boxed{\frac{\sin 4A - \sin 2A}{\cos 4A + \cos 2A} \equiv \tan A}$$

Eyeballing and Mental Gymnastics

1. $\sin 4A - \sin 2A$ *suggests* $\sin S - \sin T$
2. $\cos 4A + \cos 2A$ *suggests* $\cos S + \cos T$
3. $t = s/c$
4. *rearrange and simplify.*

$$\text{LHS} = \frac{\sin 4A - \sin 2A}{\cos 4A + \cos 2A}$$

$$= \frac{2 \cos \dfrac{4A + 2A}{2} \sin \dfrac{4A - 2A}{2}}{2 \cos \dfrac{4A + 2A}{2} \cos \dfrac{4A - 2A}{2}}$$

$$= \frac{\sin A}{\cos A}$$

$$= \tan A$$

$$\equiv \text{RHS.}$$

$$\boxed{\frac{\cos A - \cos 3A}{\sin 3A + \sin A} \equiv \tan A}$$

Eyeballing and Mental Gymnastics

1. $\cos S - \cos T$
2. $\sin S + \sin T$
3. *rearrange and simplify.*

$$\text{LHS} = \frac{\cos A - \cos 3A}{\sin 3A + \sin A}$$

$$= \frac{-2 \sin \dfrac{A + 3A}{2} \sin \dfrac{A - 3A}{2}}{2 \sin \dfrac{A + 3A}{2} \cos \dfrac{A - 3A}{2}}$$

$$= -\frac{\sin(-A)}{\cos(-A)}$$

$$= \frac{\sin A}{\cos A}$$

$$= \tan A$$

$$\equiv \text{RHS}.$$

$$\boxed{\frac{\cos A + \cos 3A}{2 \cos 2A} \equiv \cos A}$$

Eyeballing and Mental Gymnastics

1. $\cos S + \cos T$
2. *rearrange and simplify.*

$$\text{LHS} = \frac{\cos A + \cos 3A}{2 \cos A}$$

$$= \frac{2 \cos \dfrac{A + 3A}{2} \cos \dfrac{A - 3A}{2}}{2 \cos 2A}$$

$$= \frac{2 \cos 2A \cos(-A)}{2 \cos 2A}$$

$$= \cos(-A)$$

$$= \cos A$$

$$\equiv \text{RHS}.$$

$$\boxed{\frac{\sin A - \sin 3A}{\sin^2 A - \cos^2 A} \equiv 2 \sin A}$$

Eyeballing and Mental Gymnastics

1. $\sin A - \sin 3A$ *suggests* $\sin S - \sin T$ *formula*
2. $s^2 - c^2$ *suggests* $\cos 2A \equiv c^2 - s^2$
3. *rearrange and simplify.*

$$\text{LHS} = \frac{\sin A - \sin 3A}{\sin^2 A - \cos^2 A}$$

$$= \frac{2 \cos \dfrac{A + 3A}{2} \sin \dfrac{A - 3A}{2}}{- \cos 2A}$$

$$= \frac{2 \cos 2A \sin(-A)}{- \cos 2A}$$

$$= 2 \sin A$$

$$\equiv \text{RHS.}$$

$$\frac{\sin(A+B) - \sin(A-B)}{\cos(A+B) + \cos(A-B)} \equiv \tan B$$

Eyeballing and Mental Gymnastics

1. $\dfrac{\sin S - \sin T}{\cos S - \cos T}$
2. *rearrange and simplify.*

$$\text{LHS} = \frac{\sin(A+B) - \sin(A-B)}{\cos(A+B) + \cos(A-B)}$$

$$= \frac{2\cos \dfrac{(A+B)+(A-B)}{2} \sin \dfrac{(A+B)-(A-B)}{2}}{2\cos \dfrac{(A+B)+(A-B)}{2} \cos \dfrac{(A+B)-(A-B)}{2}}$$

$$= \frac{\sin B}{\cos B}$$

$$= \tan B$$

$$\equiv \text{RHS}.$$

$$\boxed{\frac{\cos A - \cos 5A}{\sin 5A + \sin A} \equiv \tan 2A}$$

Eyeballing and Mental Gymnastics

1. $\cos S - \cos T$
2. $\sin S + \sin T$
3. *rearrange and simplify.*

$$\text{LHS} = \frac{\cos A - \cos 5A}{\sin 5A + \sin A}$$

$$= \frac{-2 \sin \dfrac{A + 5A}{2} \sin \dfrac{A - 5A}{2}}{2 \sin \dfrac{A + 5A}{2} \cos \dfrac{A - 5A}{2}}$$

$$= \frac{-\sin(-2A)}{\cos(-2A)}$$

$$= \frac{\sin 2A}{\cos 2A}$$

$$= \tan 2A$$

$$\equiv \text{RHS.}$$

$$\boxed{\frac{\sin A + \sin 2A + \sin 3A}{\cos A + \cos 2A + \cos 3A} \equiv \tan 2A}$$

Eyeballing and Mental Gymnastics

1. $\sin A + \sin 2A + \sin 3A$, $\cos A + \cos 2A + \cos 3A$ *suggest* $\sin S + \sin T$ *and* $\cos S + \cos T$ *formulas using* A *and* $3A$
2. $t = s/c$
3. *rearrange and simplify.*

$$\text{LHS} = \frac{\sin A + \sin 2A + \sin 3A}{\cos A + \cos 2A + \cos 3A}$$

$$= \frac{\sin 2A + (\sin A + \sin 3A)}{\cos 2A + (\cos A + \cos 3A)}$$

$$= \frac{\sin 2A + \left(2 \sin \dfrac{A+3A}{2} \cos \dfrac{A-3A}{2}\right)}{\cos 2A + \left(2 \cos \dfrac{A+3A}{2} \cos \dfrac{A-3A}{2}\right)}$$

$$= \frac{\sin 2A + (2 \sin 2A \cos(-A))}{\cos 2A + (2 \cos 2A \cos(-A))}$$

$$= \frac{3 \sin 2A}{3 \cos 2A}$$

$$= \frac{\sin 2A}{\cos 2A}$$

$$= \tan 2A$$

$$\equiv \text{RHS.}$$

$$\frac{\sin 4A + \sin 2A}{\cos 4A + \cos 2A} \equiv \tan 3A$$

Eyeballing and Mental Gymnastics

1. $\sin S + \sin T$
2. $\cos S + \cos T$
3. *rearrange and simplify.*

$$\text{LHS} = \frac{\sin 4A + \sin 2A}{\cos 4A + \cos 2A}$$

$$= \frac{2 \sin \dfrac{4A + 2A}{2} \cos \dfrac{4A - 2A}{2}}{2 \cos \dfrac{4A + 2A}{2} \cos \dfrac{4A - 2A}{2}}$$

$$= \frac{\sin 3A \cos A}{\cos 3A \cos A}$$

$$= \tan 3A$$

$$\equiv \text{RHS.}$$

$$\boxed{\frac{\cos A - \cos 3A}{\sin 3A - \sin A} \equiv \tan 2A}$$

Eyeballing and Mental Gymnastics

1. $\cos S - \cos T$
2. $\sin S - \sin T$
3. *rearrange and simplify.*

$$\text{LHS} = \frac{\cos A - \cos 3A}{\sin 3A - \sin A}$$

$$= \frac{-2 \sin \dfrac{A + 3A}{2} \sin \dfrac{A - 3A}{2}}{2 \cos \dfrac{3A + A}{2} \sin \dfrac{3A - A}{2}}$$

$$= -\frac{\sin 2A \, \sin(-A)}{\cos 2A \, \sin A}$$

$$= \frac{\sin 2A}{\cos 2A} \cdot \frac{\sin A}{\sin A}$$

$$= \tan 2A$$

$$\equiv \text{RHS.}$$

$$\boxed{\frac{\sin 4A + \sin 8A}{\cos 4A + \cos 8A} \equiv \tan 6A}$$

Eyeballing and Mental Gymnastics

1. $\sin S + \sin T$
2. $\cos S + \cos T$
3. *rearrange and simplify.*

$$\text{LHS} = \frac{\sin 4A + \sin 8A}{\cos 4A + \cos 8A}$$

$$= \frac{2 \sin \dfrac{4A + 8A}{2} \cos \dfrac{4A - 8A}{2}}{2 \cos \dfrac{4A + 8A}{2} \cos \dfrac{4A - 8A}{2}}$$

$$= \frac{\sin 6A}{\cos 6A}$$

$$= \tan 6A$$

$$\equiv \text{RHS.}$$

$$\boxed{\dfrac{\sin 4A - \sin 8A}{\cos 4A - \cos 8A} \equiv -\cot 6A}$$

Eyeballing and Mental Gymnastics

1. $\sin S - \sin T$
2. $\cos S + \cos T$
3. *rearrange and simplify.*

$$\text{LHS} = \frac{\sin 4A - \sin 8A}{\cos 4A - \cos 8A}$$

$$= \frac{2 \cos \dfrac{4A + 8A}{2} \sin \dfrac{4A - 8A}{2}}{-2 \sin \dfrac{4A + 8A}{2} \sin \dfrac{4A - 8A}{2}}$$

$$= -\frac{\cos 6A}{\sin 6A}$$

$$= -\cot 6A$$

$$\equiv \text{RHS.}$$

$$\frac{\cos 4A - \cos 8A}{\cos 4A + \cos 8A} \equiv \tan 2A \tan 6A$$

Eyeballing and Mental Gymnastics

1. $\cos S - \cos T$
2. $\cos S + \cos T$
3. *rearrange and simplify.*

$$\text{LHS} = \frac{\cos 4A - \cos 8A}{\cos 4A + \cos 8A}$$

$$= \frac{-2 \sin \dfrac{4A + 8A}{2} \sin \dfrac{4A - 8A}{2}}{2 \cos \dfrac{4A + 8A}{2} \cos \dfrac{4A - 8A}{2}}$$

$$= -\frac{\sin 6A}{\cos 6A} \cdot \frac{\sin(-2A)}{\cos(-2A)}$$

$$= \frac{\sin 6A}{\cos 6A} \cdot \frac{\sin 2A}{\cos 2A}$$

$$= \tan 2A \tan 6A$$

$$\equiv \text{RHS.}$$

$$\frac{\sin 4A + \sin 8A}{\sin 4A - \sin 8A} \equiv \frac{-\tan 6A}{\tan 2A}$$

Eyeballing and Mental Gymnastics

1. $\sin S + \sin T$
2. $\sin S - \sin T$
3. *rearrange and simplify.*

$$\text{LHS} = \frac{\sin 4A + \sin 8A}{\sin 4A - \sin 8A}$$

$$= \frac{2 \sin \dfrac{4A + 8A}{2} \cos \dfrac{4A - 8A}{2}}{2 \cos \dfrac{4A + 8A}{2} \sin \dfrac{4A - 8A}{2}}$$

$$= \frac{\sin 6A}{\cos 6A} \cdot \frac{\cos(-2A)}{\sin(-2A)}$$

$$= \tan 6A \cdot \left(\frac{\cos 2A}{-\sin 2A} \right)$$

$$= \frac{-\tan 6A}{\tan 2A}$$

$$\equiv \text{RHS}.$$

$$\boxed{\sin 5A + 2\sin 3A + \sin A \equiv 4\sin 3A\cos^2 A}$$

Eyeballing and Mental Gymnastics

1. $\sin 5A + 2\sin 3A + \sin A$ *suggests* $\sin S + \sin T$ *formula using* $5A$ *and* A *to give* $3A$ *which is present on both LHS and RHS.*
2. *rearrange and simplify.*

$$\text{LHS} = \sin 5A + 2\sin 3A + \sin A$$

$$= (2\sin 3A) + (\sin 5A + \sin A)$$

$$= (2\sin 3A) + \left(2\sin\frac{5A+A}{2}\cos\frac{5A-A}{2}\right)$$

$$= 2\sin 3A + 2\sin 3A\cos 2A$$

$$= 2\sin 3A(1 + \cos 2A)$$

$$= 2\sin 3A(1 + (2\cos^2 A - 1))$$

$$= 2\sin 3A(2\cos^2 A)$$

$$= 4\sin 3A\cos^2 A$$

$$\equiv \text{RHS.}$$

$$1 - \cos 2A + \cos 4A - \cos 6A \equiv 4 \sin A \cos 2A \sin 3A$$

Eyeballing and Mental Gymnastics

1. $\cos 4A$, $\cos 6A$ *suggest* $\cos S - \cos T$
2. $(1 - \cos 2A)$ *suggests* $1 - \cos 2A \equiv 2 \sin^2 A$
3. *rearrange and simplify.*

$$\text{LHS} = 1 - \cos 2A + \cos 4A - \cos 6A$$

$$= (1 - \cos 2A) + (\cos 4A - \cos 6A)$$

$$= (2\sin^2 A) + \left(-2 \sin\left(\frac{4A + 6A}{2}\right) \sin\left(\frac{4A - 6A}{2}\right)\right)$$

$$= (2\sin^2 A) - (2(\sin 5A)\sin(-A))$$

$$= 2\sin^2 A + 2\sin A \sin 5A$$

$$= 2\sin A(\sin A + \sin 5A)$$

$$= 2\sin A\left(2 \sin \frac{A + 5A}{2} \cos \frac{A - 5A}{2}\right)$$

$$= 4\sin A(\sin 3A \cos(-2A))$$

$$= 4\sin A \sin 3A \cos 2A$$

$$= 4\sin A \cos 2A \sin 3A$$

$$\equiv \text{RHS}.$$

$$\boxed{\frac{\cos 2A - \cos 4A}{\cos 2A + \cos 4A} - \tan 3A \tan A \equiv 0}$$

Eyeballing and Mental Gymnastics

1. $\cos S - \cos T$
2. $\cos S + \cos T$
3. $t = s/c$
4. *rearrange and simplify.*

$$\text{LHS} = \frac{\cos 2A - \cos 4A}{\cos 2A + \cos 4A} - \tan 3A \tan A$$

$$= \frac{-2 \sin\left(\dfrac{2A + 4A}{2}\right) \sin\left(\dfrac{2A - 4A}{2}\right)}{2 \cos\left(\dfrac{2A + 4A}{2}\right) \cos\left(\dfrac{2A - 4A}{2}\right)} - \tan 3A \tan A$$

$$= -\frac{\sin 3A}{\cos 3A} \cdot \frac{\sin(-A)}{\cos(-A)} - \tan 3A \tan A$$

$$= \tan 3A \tan A - \tan 3A \tan A$$

$$= 0$$

$$\equiv \text{RHS.}$$

$$\boxed{\frac{\sin 2A + \sin 4A}{\sin 2A - \sin 4A} + \frac{\tan 3A}{\tan A} \equiv 0}$$

Eyeballing and Mental Gymnastics

1. $\sin S + \sin T$
2. $\sin S - \sin T$
3. $t = s/c$
4. *rearrange and simplify.*

$$\text{LHS} = \frac{\sin 2A + \sin 4A}{\sin 2A - \sin 4A} + \frac{\tan 3A}{\tan A}$$

$$= \frac{2 \sin\left(\dfrac{2A + 4A}{2}\right) \cos\left(\dfrac{2A - 4A}{2}\right)}{2 \cos\left(\dfrac{2A + 4A}{2}\right) \sin\left(\dfrac{2A - 4A}{2}\right)} + \frac{\tan 3A}{\tan A}$$

$$= \frac{\sin 3A}{\cos 3A} \cdot \frac{\cos(-A)}{\sin(-A)} + \frac{\tan 3A}{\tan A}$$

$$= \tan 3A \cdot \left(-\frac{1}{\tan A}\right) + \frac{\tan 3A}{\tan A}$$

$$= -\frac{\tan 3A}{\tan A} + \frac{\tan 3A}{\tan A}$$

$$= 0$$

$$\equiv \text{RHS.}$$

$$\boxed{\cos 2A - \cos 10A \equiv \tan 4A(\sin 2A + \sin 10A)}$$

Eyeballing and Mental Gymnastics

1. $(\cos 2A - \cos 10A)$ *suggest* $\cos S - \cos T$
2. $(\sin 2A + \sin 10A)$ *suggests* $\sin S + \sin T$
3. *both sides are complex;* $\therefore$ *may have to simplify both sides*
4. $\tan 4A$ *may come from* $(10A - 2A)/2$
5. *rearrange and simplify.*

$$\text{LHS} = \cos 2A - \cos 10A$$

$$= -2 \sin \left(\frac{2A + 10}{2} \right) \sin \left(\frac{2A - 10A}{2} \right)$$

$$= -2 \sin 6A \sin(-4A)$$

$$= 2 \sin 6A \sin 4A.$$

$$\text{RHS} = \tan 4A(\sin 2A + \sin 10A)$$

$$= \tan 4A \left(2 \sin \left(\frac{2A + 10}{2} \right) \cos \left(\frac{2A - 10A}{2} \right) \right)$$

$$= \tan 4A(2 \sin 6A) \cos(-4A))$$

$$= \tan 4A \cdot 2 \sin 6A \cos 4A$$

$$= \frac{\sin 4A}{\cos 4A} \cdot 2 \sin 6A \cos 4A$$

$$= 2 \sin 6A \sin 4A$$

$$\therefore \text{ LHS} \equiv \text{RHS}.$$

$$\boxed{\sin 2A + \sin 4A - \sin 6A \equiv 4 \sin A \sin 2A \sin 3A}$$

Eyeballing and Mental Gymnastics

1. $\sin 2A - \sin 6A$ *suggests* $\sin S - \sin T$ *formula*
2. $\sin 4A$ *suggests " double angle" formula to give* $2A$
3. *rearrange and simplify.*

$$\text{LHS} = \sin 2A + (\sin 4A) - \sin 6A$$

$$= 2 \cos \frac{2A + 6A}{2} \sin \frac{2A - 6A}{2} + (\sin 4A)$$

$$= 2 \cos 4A \sin(-2A) + \sin 4A$$

$$= -2 \cos 4A \sin 2A + 2 \sin 2A \cos 2A$$

$$= 2 \sin 2A(\cos 2A - \cos 4A)$$

$$= 2 \sin 2A \left(-2 \sin \frac{2A + 4A}{2} \sin \frac{2A - 4A}{2} \right)$$

$$= 2 \sin 2A(-2 \sin 3A \sin(-A))$$

$$= 2 \sin 2A(2 \sin 3A \sin A)$$

$$= 4 \sin A \sin 2A \sin 3A$$

$$\equiv \text{RHS}.$$

$$\boxed{\cos 3A + 2 \cos 5A + \cos 7A \equiv 4 \cos^2 A \cos 5A}$$

Eyeballing and Mental Gymnastics

1. $\cos 3A$, $\cos 7A$ suggest $\cos S + \cos T$ *formula*
2. *rearrange and simplify.*

$$\text{LHS} = \cos 3A + 2 \cos 5A + \cos 7A$$

$$= (\cos 3A + \cos 7A) + 2 \cos 5A$$

$$= \left(2 \cos \frac{3A + 7A}{2} \cos \frac{7A - 3A}{2} \right) + 2 \cos 5A$$

$$= (2 \cos 5A \cos 2A) + 2 \cos 5A$$

$$= 2 \cos 5A (\cos 2A + 1)$$

$$= 2 \cos 5A (2 \cos^2 A - 1 + 1)$$

$$= 4 \cos^2 A \cos 5A$$

$$\equiv \text{RHS.}$$

$$1 + \cos 2A + \cos 4A + \cos 6A \equiv 4 \cos A \cos 2A \cos 3A$$

Eyeballing and Mental Gymnastics

1. $\cos 2A$, $\cos 4A$ *suggest* $\cos S + \cos T$ *formula*
2. $\cos 6A$ *suggests* $\cos 2(3A)$
3. *rearrange and simplify.*

$$\text{LHS} = 1 + \cos 2A + \cos 4A + \cos 6A$$

$$= 1 + 2 \cos \frac{2A + 4A}{2} \cos \frac{2A - 4A}{2} + \cos 6A$$

$$= 1 + 2 \cos 3A \cos(-A) + \cos 6A$$

$$= 1 + 2 \cos 3A \cos A + \cos 2(3A)$$

$$= 1 + 2 \cos 3A \cos A + (2 \cos^2 3A - 1)$$

$$= 2 \cos 3A \cos A + 2 \cos^2 3A$$

$$= 2 \cos 3A (\cos A + \cos 3A)$$

$$= 2 \cos 3A \left(2 \cos \frac{A + 3A}{2} \cdot \cos \frac{A - 3A}{2} \right)$$

$$= 4 \cos 3A \cos 2A \cos(-A)$$

$$= 4 \cos A \cos 2A \cos 3A$$

$$\equiv \text{RHS}.$$

$$1 - \cos 2A + \cos 4A - \cos 6A \equiv 4 \sin A \cos 2A \sin 3A$$

Eyeballing and Mental Gymnastics

1. $\cos 2A$, $\cos 4A$ suggest $\cos S - \cos T$
2. $\cos 6A$ suggests $\cos 2(3A)$
3. *rearrange and simplify.*

$$\text{LHS} = 1 - \cos 2A + \cos 4A - \cos 6A$$

$$= 1 + (\cos 4A - \cos 2A) - \cos 6A$$

$$= 1 + \left(-2 \sin \frac{4A + 2A}{2} \sin \frac{4A - 2A}{2}\right) - \cos 6A$$

$$= 1 - 2 \sin 3A \sin A - \cos 2(3A)$$

$$= 1 - 2 \sin 3A \sin A - (1 - 2 \sin^2 3A)$$

$$= -2 \sin 3A \sin A + 2 \sin^2 3A$$

$$= 2 \sin 3A (\sin 3A - \sin A)$$

$$= 2 \sin 3A \left(2 \cos \frac{3A + A}{2} \sin \frac{3A - A}{2}\right)$$

$$= 2 \sin 3A (2 \cos 2A \sin A)$$

$$= 4 \sin A \cos 2A \sin 3A$$

$$\equiv \text{RHS.}$$

$$\tan 4A(\sin 2A + \sin 10A) \equiv \cos 2A - \cos 10A$$

Eyeballing and Mental Gymnastics

1. $t = s/c$
2. $\sin S + \sin T$
3. $\cos S - \cos T$
4. *simplify both sides before comparison.*

$$\text{LHS} = \tan 4A(\sin 2A + \sin 10A)$$

$$= \tan 4A \left(2 \sin \left(\frac{2A + 10A}{2} \right) \cos \left(\frac{2A - 10A}{2} \right) \right)$$

$$= \frac{\sin 4A}{\cos 4A} (2 \sin 6A \cos(-4A))$$

$$= \frac{\sin 4A}{\cos 4A} 2 \sin 6A \cos 4A$$

$$= 2 \sin 6A \sin 4A$$

$$\text{RHS} = \cos 2A - \cos 10A$$

$$= -2 \sin \left(\frac{2A + 10A}{2} \right) \sin \left(\frac{2A - 10A}{2} \right)$$

$$= -2 \sin 6A \sin(-4A)$$

$$= 2 \sin 6A \sin 4A$$

$$\text{LHS} \equiv \text{RHS}.$$

$$\boxed{\sin A(\sin 3A + \sin 5A) \equiv \cos A(\cos 3A - \cos 5A)}$$

Eyeballing and Mental Gymnastics

1. *Both sides are complex; therefore it may be easier to simplify both sides before comparison*
2. $\sin S + \sin T$
3. $\cos S - \cos T$
4. *rearrange and simplify.*

$$\text{LHS} = \sin A(\sin 3A + \sin 5A)$$

$$= \sin A \left(2 \sin \frac{3A + 5A}{2} \cos \frac{3A - 5A}{2} \right)$$

$$= \sin A(2 \sin 4A \cos(-A))$$

$$= 2 \sin A \cos A \sin 4A$$

$$\text{RHS} = \cos A(\cos 3A - \cos 5A)$$

$$= \cos A \left(-2 \sin \frac{3A + 5A}{2} \sin \frac{3A - 5A}{2} \right)$$

$$= \cos A(-2 \sin 4A \sin(-A))$$

$$= 2 \sin A \cos A \sin 4A$$

$$\text{LHS} \equiv \text{RHS}.$$

$$\boxed{\sin A(\sin A + \sin 3A) \equiv \cos A(\cos A - \cos 3A)}$$

Eyeballing and Mental Gymnastics

1. *Both sides appear to be complex and therefore there may be a need to simplify both sides before comparison is made*
2. $\sin S + \sin T$
3. $\cos S - \cos T$
4. *rearrange and simplify.*

$$\text{LHS} = \sin A(\sin A + \sin 3A)$$

$$= \sin A\left(2\sin\frac{A+3A}{2}\cos\frac{A-3A}{2}\right)$$

$$= \sin A(2\sin 2A\cos(-A))$$

$$= 2\sin A\cos A\sin 2A$$

$$\text{RHS} = \cos A(\cos A - \cos 3A)$$

$$= \cos A\left(-2\sin\frac{A+3A}{2}\sin\frac{A-3A}{2}\right)$$

$$= \cos A(-2\sin 2A\sin(-A))$$

$$= 2\sin A\cos A\sin 2A$$

$$\text{LHS} \equiv \text{RHS}.$$

$$\boxed{\tan A + \tan(A + 120°) + \tan(A + 240°) \equiv 3\tan 3A}^{*}$$

Eyeballing and Mental Gymnastics

1. $\tan(A + 120°)$, $\tan(A + 240°)$, $\tan 3A$ *suggest expansion of "compound angles"*
2. $\tan 120° = \tan(180° - 60°) = -\tan 60° = -\sqrt{3}$
3. $\tan 240° = \tan(180° + 60°) = \tan 60° = \sqrt{3}$
4. *both LHS and RHS are complex; therefore may have to expand both sides for ease of comparison*
5. $\tan 3A$ *suggest* $\tan(2A + A)$ *and* $\tan 2A$ *expansions*
6. *rearrange and simplify.*

$$\text{LHS} = \tan A + \tan(A + 120°) + \tan(A + 240°)$$

$$= \tan A + \left(\frac{\tan A + \tan 120°}{1 - \tan A \tan 120°} \right) + \left(\frac{\tan A + \tan 240°}{1 - \tan A \tan 240°} \right)$$

$$= \tan A + \left(\frac{\tan A - \sqrt{3}}{1 + \sqrt{3}\tan A} \right) + \left(\frac{\tan A + \sqrt{3}}{1 - \sqrt{3}\tan A} \right)$$

$$= \tan A + \frac{(\tan A - \sqrt{3})(1 - \sqrt{3}\tan A) + (\tan A + \sqrt{3})(1 + \sqrt{3}\tan A)}{(1 + \sqrt{3}\tan A)(1 - \sqrt{3}\tan A)}$$

$$= \tan A + [(\tan A - \sqrt{3}\tan^2 A - \sqrt{3} + 3\tan A)$$
$$+ (\tan A + \sqrt{3}\tan^2 A + \sqrt{3} + 3\tan A)]/(1 - 3\tan^2 A)$$

$$= \tan A + \frac{8\tan A}{1 - 3\tan^2 A}$$

$$= \frac{\tan A(1 - 3\tan^2 A) + 8\tan A}{1 - 3\tan^2 A}$$

$$= \frac{9\tan A - 3\tan^3 A}{1 - 3\tan^2 A}$$

$$= \frac{3(3\tan A - \tan^3 A)}{1 - 3\tan^2 A}$$

$$\text{RHS} = 3 \tan 3A$$

$$= 3 \tan(A + 2A)$$

$$= 3 \left(\frac{\tan A + \tan 2A}{1 - \tan A \tan 2A} \right)$$

$$= 3 \left(\frac{\tan A + \left(\dfrac{2 \tan A}{1 - \tan^2 A} \right)}{1 - \tan A \left(\dfrac{2 \tan A}{1 - \tan^2 A} \right)} \right)$$

$$= 3 \left(\frac{\dfrac{\tan A(1 - \tan^2 A) + 2 \tan A}{1 - \tan^2 A}}{\dfrac{(1 - \tan^2 A) - 2 \tan^2 A}{1 - \tan^2 A}} \right)$$

$$= 3 \left(\frac{\tan A - \tan^3 A + 2 \tan A}{1 - \tan^2 A - 2 \tan^2 A} \right)$$

$$= \frac{3(3 \tan A - \tan^3 A)}{1 - 3 \tan^2 A}$$

$$\therefore \text{LHS} \equiv \text{RHS}.$$

An alternative proof is to write the second and third terms of the LHS in terms of sin/cos; then add them using common denominator, followed by "compound angle formula" for the numerator, and "cos S + cos T formula" for the denominator. This second proof is shorter and more elegant. Try it out for yourself.

*Note that there is greater beauty if the identity is written as:

$$\tan A + \tan(A + 120°) + \tan(A + 240°) \equiv 3 \tan(A + 360°).$$

Of course, $\tan(A + 360°) = \tan A$.

Addenda

Proof for sin(*A* + *B*)

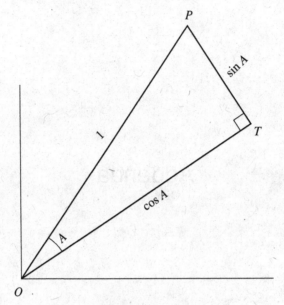

Fig. 1

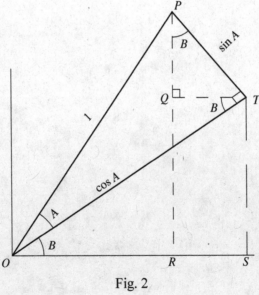

Fig. 2

Proof for sin(*A* + *B*)

Let us draw a right-angled triangle with angle *A* as in Fig. 1 (opposite page)

then : $\sin A = O/H = \sin A/1$

$\cos A = A/H = \cos A/1$

Let's add a second triangle with angle B to the first triangle. Figure 2 (opposite page)

now : $\sin(A+B) = \dfrac{O}{H} = \dfrac{PR}{1}$

$= PQ + QR$ (from Fig. 2)

From geometry, we know that $\angle QTO = B$ (alt $\angle$'s between // lines)
Similarly, $\angle QPT = B$ (both are complementary $\angle$s of $\angle PTQ$)

For triangle PQT,

$$\cos B = \frac{PQ}{PT} = \frac{PQ}{\sin A}$$

$$\therefore PQ = \sin A \cos B$$

For triangle TOS,

$$\sin B = \frac{TS}{OT} = \frac{QR}{\cos A}$$

$$\therefore QR = \cos A \sin B$$

$$\sin(A+B) = PQ + QR$$

$$= \sin A \cos B + \cos A \sin B$$

$$\therefore \sin(A+B) \equiv \sin A \cos B + \cos A \sin B$$

Proof for cos(*A* + *B*)

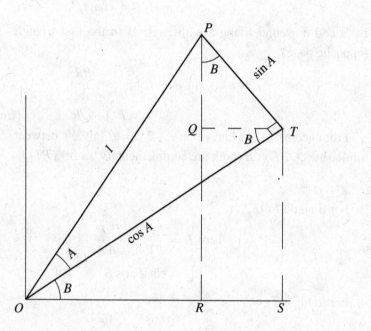

Proof for cos(A + B)

From the Fig (opposite page)

$$\cos(A+B) = \frac{OR}{OP} = \frac{OR}{1}$$

$$= OS - RS$$

$\angle QTO = B$ (alt $\angle$'s between $//$ lines)

$\angle QPT = \angle QTO$ (both are complementary angles to $\angle PTQ$)

$$= B$$

For triangle PQT:

$$\sin B = \frac{QT}{PT} = \frac{QT}{\sin A}$$

$$\therefore QT = \sin A \sin B$$

For triangle TOS;

$$\cos B = \frac{OS}{OT} = \frac{OS}{\cos A}$$

$$\therefore OS = \cos A \cos B$$

$$\cos(A+B) = OS - RS$$

$$= OS - QT \qquad \text{(since RS=QT)}$$

$$= \cos A \cos B - \sin A \sin B$$

$$\therefore \cos(A+B) \equiv \cos A \cos B - \sin A \sin B$$

Proof for tan(A + B)

$$\tan(A+B) = \frac{\sin(A+B)}{\cos(A+B)}$$

$$= \frac{\sin A \cos B + \cos A \sin B}{\cos A \cos B - \sin A \sin B}$$

$$= \frac{\tan A + \tan B}{1 - \tan A \tan B} \qquad \left|\begin{array}{l}\text{divide all}\\ \text{four terms by}\\ \cos A \cos B\end{array}\right.$$

Substitute B by $(-B)$

$$\tan(A + (-B)) = \frac{\tan A + \tan(-B)}{1 - \tan A \tan(-B)}$$

$$\therefore \ \tan(A - B) = \frac{\tan A - \tan B}{1 + \tan A \tan B}$$

Since $\tan(-B) = -\tan B$.

Appetisers for Higher Trigonometry

$$\sin x = x - \frac{x^3}{3!} + \frac{x^5}{5!} - \frac{x^7}{7!} + \frac{x^9}{9!} - \cdots$$

$$\cos x = 1 - \frac{x^2}{2!} + \frac{x^4}{4!} - \frac{x^6}{6!} + \frac{x^8}{8!} - \cdots$$

$$\tan x = x + \frac{x^3}{3} + \frac{2x^5}{15} + \cdots$$

$$\arcsin x = x - \frac{1}{2}\frac{x^3}{3} + \frac{1\cdot3}{2\cdot4}\frac{x^5}{5} - \frac{1\cdot3\cdot5}{2\cdot4\cdot6}\cdot\frac{x^7}{7} + \cdots$$

$$\arctan x = x - \frac{x^3}{3} + \frac{x^5}{5} - \frac{x^7}{7} + \cdots$$

$$e^x = 1 + \frac{x}{1!} + \frac{x^2}{2!} + \frac{x^3}{3!} + \frac{x^4}{4!} + \cdots$$

$$e^{ix} = 1 + \frac{(ix)}{1!} + \frac{(ix)^2}{2!} + \frac{(ix)^3}{3!} + \frac{(ix)^4}{4!} + \cdots$$

$$= \begin{cases} 1 - \dfrac{x^2}{2!} + \dfrac{x^4}{4!} - \dfrac{x^6}{6!} + \dfrac{x^8}{8!} - \cdots \\[2mm] + \dfrac{ix}{1!} - \dfrac{ix^3}{3!} + \dfrac{ix^5}{5!} - \cdots \end{cases}$$

$$e^{ix} = \cos x + i\sin x$$

$$e^{i\pi} = \cos \pi + i\sin \pi$$

$$= (-1) + i(0)$$

$$e^{i\pi} = -1$$

About the Author

Dr Y E O Adrian graduated from the University of Singapore with first class Honours in Chemistry in 1966, and followed up with a Master of Science degree in 1968.

He received his Master of Arts and his Doctor of Philosophy degrees from Cambridge University in 1970, and did post-doctoral research at Stanford University, California.

For his research, he was elected Fellow of Christ's College, Cambridge and appointed Research Associate at Stanford University in 1970.

His career spans fundamental and applied research and development, academia, and top appointments in politics and industry. His public service includes philanthropy and sports administration. Among his numerous awards are the Charles Darwin Memorial Prize, the Republic of Singapore's Distinguished Service Order, the International Olympic Committee Centenary Medal, and the Honorary Fellowship of Christ's College, Cambridge University.